C000254048

THE
PIT SINKERS
OF
NORTHUMBERLAND
AND DURHAM

To Lamy

November 2012

Pit sinkers at Boldon Colliery, Durham, in 1866. (Boldon Historical Society)

THE PIT SINKERS

OF

NORTHUMBERLAND AND DURHAM

PETER FORD MASON

The History Press

In memory of my daughter, Sandra

Front Cover: Sinkers descending in kibbles at Marsden Colliery, Durham, in 1881.
(Durham Record Office)

First published 2012

The History Press
The Mill, Brimscombe Port
Stroud, Gloucestershire, GL5 2QG
www.thehistorypress.co.uk

© Peter Ford Mason, 2012

The right of Peter Ford Mason to be identified as the Author
of this work has been asserted in accordance with the
Copyrights, Designs and Patents Act 1988.

All rights reserved. No part of this book may be reprinted
or reproduced or utilised in any form or by any electronic,
mechanical or other means, now known or hereafter invented,
including photocopying and recording, or in any information
storage or retrieval system, without the permission in writing
from the Publishers.

British Library Cataloguing in Publication Data.
A catalogue record for this book is available from the British Library.

ISBN 978 0 7524 8094 7

Typesetting and origination by The History Press
Printed in Great Britain

Contents

Acknowledgements

My earliest introduction into writing outside formal education came from my mother, who encouraged my twin brother and I to delve into the mysteries of my father's family tree. I wish to thank my family for their continuing support, especially my son David who has provided valuable IT assistance. I also wish to give credit and thanks to the following organisations which have helped in the final stages towards publication:

Beamish Museum
Cornwall Record Office
Derbyshire Record Office
Durham City Library
The Durham Mining Museum
Durham Record Office
Flintshire Record Office
Glamorgan Record Office
Heswall Library
Institution of Civil Engineers
Lancashire Record Office
Leicester Record Office
Manchester Art Gallery Picture Library
The National Archives
National Coal Museum
National Museums, Liverpool
Neston Library
Newcastle City Library
The North of England Institute of Mining and Mechanical Engineers
Northumberland Record Office
Nottinghamshire Record Office
Scottish Record Office
Shropshire Record Office
Somerset Record Office
Staffordshire Record Office

Introduction

Irish Professor John Tyndal, former civil and railway engineer, stated in 1866:

> Wherever two atoms capable of uniting together by their mutual attractions exit separately, they form a store of potential energy. Thus our woods, forests, and coal-fields on the one hand, and our atmospheric oxygen on the other, constitute a vast store of energy, of this kind – vast, but far from infinite.

J.L. and B. Hammond, social historians, in their book *The Town Labourer 1760–1832* noted that:

> A civilisation is the use to which an age puts it resources of wealth, knowledge and power, in order to create a social life.

Memorial in Durham Cathedral:

> They break out a shaft away from where men sojourn,
> they are forgotten of the foot that passeth by.
>
> Book of Job

★ ★ ★

The need to burrow into the ground has always been a human and animal instinct in order to provide security from predators, to gain resources and to bury their dead. For cave dwellers, holes were dug as animal traps and for finding water. Excavations have also been used to secure underground storage, whether it was squirrels hiding away their nuts or nations seeking places for waste material. Activities as diverse as potholing and exploratory boring through the Earth's crust show that the subterranean world continues to intrigue us.

Also from early times man's tunnelling and mining skills have been put to use when attacking castles and strongholds or escaping from prisons and concentration camps. Mining engineers and civil engineers took over from military engineers in the sinking of shafts (vertical tunnels) to obtain minerals and coal, and in the construction of canal, railway, road and underwater tunnels. The words miner, minerals and coal are closely associated, although coal does not strictly qualify as a mineral since it was made from organic substances. Sinking described the descent to 'win' those materials hidden underground. This work was carried out by a 'sinker'.

Particularly in coal mining, the need for adequate ventilation became a pressing matter when explosive gases were encountered as shafts and mining went deeper. From the late eighteenth century, the human death toll has been staggering due to the increasingly large numbers of men and boys involved in mining, and the reluctance and short-sightedness of many mine owners to improve working conditions. The sinkers were exposed to great dangers as they perilously worked in the shafts. In early times they were simply held on the end of a rope and hook and their lives depended on these – hook, line and sinker!

The fireplace has been the focal point of the home, giving warmth and security, whether a stately mansion or a lowly cottage. However, the fuel which started as wood or peat, and moved on to coal, has now has been taken over by gas and other more environmentally friendly resources, leading to the decline of coal mining.

This book is dedicated to those mining communities who gave so much during the Industrial Revolution and to their struggle to obtain basic human rights which have come to benefit all of us in following generations. The miners provided a 'shaft of light' for the struggling working classes during a very bleak period.

1

History of Shaft Sinking

The sinking of pits and shafts, and the subsequent mining, were carried out in a simple manner until the advent of steam power dramatically changed the scale of underground operations, but not always to the benefit of the sinker and miner.

Up to Agricola (1540)

Mining may well have been the second of mankind's earliest endeavours after agriculture, and ranks as one of the primary industries of human civilisation. The use of wells to obtain water was one of the first examples of sinking and probably came in advance of the search for minerals. Horizontal shafting or tunnelling has always been an integral part of mining, as well as an important part of early warfare.

The Greek word *Metal'lum* signified a pit or cave, where 'anything was sought for by digging', hence any mineral found in a mine. In the Nubian Desert of northern Sudan and in the Timna Valley (now in Israel), Egyptian miners excavated circular shafts up to 90ft deep. The excavation tools used were primitive, consisting of metal chisels and hoes for excavating circular shafts, with footholds in the shaft walls for moving up and down. Roman shafts were square and small (up to 6ft wide), braced with wood to prevent collapse, and could be as deep as 600ft. The Romans were also skilled tunnel builders as displayed in their civil engineering works. Their shaft sinking methods were similar to the Egyptians', using divided shafts for purposes of ventilation. Apparently coal mining took place in China as early as 200 BC, and at this time Hannibal was employing Spanish miners to make a road across the Alps.

The early miners preferred open-cast working as the easiest way of getting to the minerals or coal, but they also dug 'bell pits' (shallow shafts that widened out at the base in the shape of a bell). In Durham, these shallow pits were often worked by monks, who were pioneers in mining. They obtained considerable income from the mining of coal, and their detailed records included the sinking of pits. Monks were credited with introducing coal as fuel. Records in the year 852 noted that the Abbey of Peterborough received twelve cartloads 'of fossil or pit-coal'. Monks in various Scottish locations were digging bell pits and open-cut mines as early as the twelfth century. Bell pits were also excavated west of Newcastle where the seams were near the surface. When coal mining was not under the authority of the monasteries it was controlled by the Crown who leased the work out to those in favour.

Marco Polo, on his travels to the interior of Cathay towards the end of the thirteenth century, was surprised at the practice of the natives who burned a strange fuel which they called 'black stone ... dug out of the mountains where it runs in veins'. This coal which outcropped at the surface in many places between the rocky strata had probably been dug by the Chinese for centuries. The early Germans gave it the name 'day' (tage) coal, that is 'daylight' mining or workings at the surface, later used in Britain. One of the earliest references to coal mining in Britain was contained in Bishop Pudsey's Boldon Buke (AD 1183) describing 'a certain collier' ('carbonarius') at Escomb, County Durham, who was obliged to find coals, but this was probably referring to charcoal burning. Other early mentions of coal by men of letters included the following:

Chaucer (b.1343), *Canterbury Tales*:
    Sered pokettes, sal peter, and vitriole:
    And Divers fies made of wode and cole…

Shakespeare (b.1564):
    Thou didst swear to me ... sitting by a sea-coal fire…

Milton (b.1608):
    Or coal Tyne, or ancient hallowed Dee…

From earliest times for common people, coal was not readily available, and therefore wood was the first choice for fuel. Ancient laws allowed the use of wood for the home fire. Charcoal rather than coal was used to smelt iron; this was initially made from animal bones and not wood. In 1228, a pathway in London was known as 'Sacoles Lane' (Sea Coals Lane). Coal was occasionally outlawed by statute in London due to its polluting smoke, and there was a dislike for the new taste of food cooked using coal. In Britain, as wood supplies became scarce, there was a greater need to find more coal resources for fuel and shipbuilding. This led to drift mining, also known as adit mining. Such mining was practised at Lumley in County Durham, where the miners followed the coal seam further underground until the working conditions became unsafe, and then the mines were abandoned. At Gateshead, some drift mining preceded bell pits.

Coal was found near the surface in the West Midlands, Somerset, Leicestershire, Nottinghamshire, Yorkshire, Lancashire, North and South Wales, Northumberland, Durham and on both sides of the Firth of Forth. Often the use of bell pits was little better than pillage in the wasteful and haphazard manner it was carried out, and succeeded in turning the Black Country of southern Staffordshire into what was called a 'water logged rabbit warren', a desert of water-filled pits and smouldering waste heaps. Mining in north-east Britain started with the Romans burning and excavating coal in the region. Initially it seemed that they only used coal for making toys and bracelets, however, half-burnt cinders and the ashes of coal fires have been discovered amongst the mortar of Roman buildings near Newcastle. Up to the eighteenth century, coal outcrops were mined where coal appeared at the surface, often along river banks which afforded easy access for exploitation, probably by landowners who extracted coal for their own domestic fires. Coastal erosion caused much of the outcropping coal to be washed ashore and was occasionally collected on the Durham beaches. Newcastle became known as 'the Eye of the North and the Hearth that warmeth the South parts of this kingdom with Fire'. Newcastle was also a 'shield and defence against invasions' and it became a busy centre of industry and later the 'outlet for a vast amount of steam-power, which was exported as coal to all parts of the world'.

King Stephen granted mineral rights to the Bishop of Durham to pay for the upkeep of the cathedral and to maintain troops to protect against Scottish raids. A century later in 1239, Henry III granted a charter recognising the importance of coal supplies to the freemen of Newcastle upon Tyne, allowing them to dig for coals 'unhindered'. As early as 1256, jurors dealing with minor criminal offences were complaining that the road from Corbridge to Newcastle was dangerous *per fossas et mineras* [due to digging out of minerals], especially overnight. Scotland was also starting to mine coal, with the charter of 1291 allowing the monks of Dunfermline the privilege to dig coal in the neighbourhood of their monastery. In 1302 the inhabitants of the Palatinate of Durham petitioned the King on the rights of freemen to mine coal in their own land. Durham's Prince Bishops owned rights to the mining of coal and iron, and Durham monks owned and leased mines at Lumley. By 1334, Newcastle was the fourth-wealthiest town in England based on the coal trade, including the export of coal to many parts of England.

Coal diggers borrowed their early mining techniques and methods from the metallurgical miners. These methods to extract coal were mentioned in a Durham bailiff's account of the manor of Coundon in 1348–50 which involved the mining of sea coal 'with cords, buckets, and windlasses'. Costs were recorded as: 5s 6d including ropes, scopes and windlass; a new pit taking less than a year to work out, cost 2s 6d to 5s to sink; ropes cost 2s 6d each, skips or buckets 4d each, windlasses 2d each, and pick sharpening 12d per year. The men were paid extra for drawing or removing water. Also in 1354, Thomas de Fery leased to the Prior of Durham for thirty years, all his coals and seams of coal in the north part of the 'vill of Ferryhill', with license to dig in any place and to construct 'water gates' (adits) to control pit drainage. There was also an early reference to a joint venture between several Durham colliery owners to construct an adit 'in the lands of Hett' in 1407.

The shallow holes and quarries began to be taken over by mine works consisting of pit and adit, or vertical shaft and horizontal gallery. The latter methods were adequate providing coal workings were carried out above the level of free drainage. The shaft was used to raise coal by means of a manual windlass or jack roll, and the adit served to drain the water from the workings. Combining the two produced a natural ventilation sufficient for the shallow and limited workings at this time. In east Scotland the use of the windlass was unknown, with coals being carried up stairs or ladders in the shafts on the backs of women known as 'coal bearers'. In early mining practice, drainage was the one technical problem which was likely to have entailed substantial capital expenditure. In 1486–7 the monks of Finchale spent £99 15s 6d on a horse-driven pump, while in Nottinghamshire, the Willoughbys of Wollaton spent the considerable sum of £1,000 on driving a mile-long 'sough' (open drain), otherwise the alternative would have been the sinking of many ventilation shafts. It was only when shafts became deeper that drainage by gravity was not possible, and the natural ventilation became more feeble and irregular. Noxious gases began to imperil the safety of miners. Dr Kaye, founder of Caius College, Cambridge, noted that in certain northern coal pits 'the unwholesome vapour whereof is so pernicious to the hired labourers that it would immediately destroy them if they did not get out of the way as soon as the flame of their lamps becomes blue and is consumed'. Explosions in German mines were thought to be the breath of demons!

Flooding was another risk that confronted early miners. In 1515 a flooded mine at Liege in Belgium caused the deaths of eighty-eight miners. An accident of this magnitude could hardly have occurred in England at this time; for until the days of Queen Elizabeth I the number of miners working together in a single pit or drift probably did not exceed a dozen or so, and the colliery owner who kept more than three pits in operation was quite exceptional. The change from woodland to arable land in Britain, causing a dire lack of wood, was a great incentive to find methods to increase the production of coal. The Elizabethan reign marked the beginning of an epoch in the history of coal mining as new methods pioneered by mineralogist Agricola were introduced from abroad.

From Agricola to start of steam power (1540–1700)

Professor Neff has supported the view that during the period 1540–1640 the first Industrial Revolution took place. Especially with iron-making, the workers were dependent upon an employer for their raw materials and market. They were brought together in a 'works', and were paid wages. It would be some time before the coal industry would organise itself on similar lines. The change was to start in the Newcastle area where shipments of coal increased at a more rapid rate than at any period in its history, and this was based on mining at deeper levels.

Records show that in Elizabethan Tynemouth there were small mines in town-fields. These were pits 5 fathoms deep, which took twelve days to sink, cost £2 each, and produced 38 tons a day. Tyneside was to have the advantage due to its superior location near navigable water. In Germany, lack of this benefit held back the development of the Ruhr and Saar coalfields. In 1564, German miners under the guidance of mining engineer and entrepreneur Daniel Hechstetter, were invited to Britain by Elizabeth to search for and extract all kinds of minerals. These Germans were considered the finest miners at the cutting edge of technology. Their mining techniques, mainly based on drift or outcrop-type mining, were well catalogued by Georgius Agricola, 'Father of Mining', c.1556, in his book *De Re Metallica* (Latin for 'On the Nature of Metals' [meaning minerals]).

In Durham and Northumberland, the Crown employed certain permanent officials to supervise the mining of coal in the royal domains. In 1557 William Dixson was 'deputie viewer of the King's cole mynes in Benwell' and in 1581 Edward Bulmer was appointed 'viewer of the Queen's mines in and near Newcastle'. Although most of the miners that were sought by rival colliery owners were native to Britain, there were a few exceptions: Symon de Harmsetrange of French extraction, a sinker at Gateshead in 1548, and a Thomas Gambeskie of Polish extraction, an overman at Whickham in 1609. The 'Grand Lease', allowing mining rights and privileges covering the manors of Gateshead and Whickham in Durham, was procured by the Earl of Leicester from Queen Elizabeth in 1582. These rights were transferred to the 'Society of Free Hosts' of Newcastle which gave it the monopoly of the coal trade for the next hundred years. The importance of coal was such that even when the price of coal rose, the pleadings of the Mayor of London to such a powerful figure as Lord Burleigh came to nothing.

Owing to the increase in the number of miners working in a single pit, the management of these pits came to require more than one official. The coal owners in the North East were the first to recognise the need for proper supervision of their workers. In 1582, at Winlaton Colliery, the function of a viewer or general supervisor was 'to see [the] workemen gotten to worke … and to see the ground upholden and truely wroughte underground'. They also had an interest to ensure that wastage was kept to a minimum. With the sinking of deeper shafts and the driving of longer and more numerous

Portrait of Georgius Agricola, 'Father of Mining', c.1556. (Science Photo)

headways, the task of supervising all the miners in one pit became a full-time job, and there was the need for a special underground foreman at each pit. In the Tyne valley he was called an 'under overman', in Nottingham an 'underman', and in Scotland an 'oversman'.

The two main factors which determined whether a pit should be sunk to extract coal for shipment were the distance of the shaft head from the navigable part of the river and the depth of the seam below the surface. Another factor which gave stimulus to coal mining in the North East was the establishment of the salt industry on the Wear in 1580, which attracted the attention of English merchants to the possibility of shipping large quantities of coal needed in that industry. Neff noted: 'Some time before 1591, Robert Bowes, Treasurer of Berwick, erected pans at Sunderland for boiling down sea water, and with a view to supplying them with fuel, invested large sums of capital, including £2,000 for an adit in a colliery, and hence the coal was brought down the Wear in keels to Sunderland.' Unlike Royalist Newcastle, due to its support of the Parliamentary cause Sunderland's trade prospered, helped by a large increase in demand for coal from the continent.

Some coal owners searched for other methods of raising the coal. An engine borrowed from the East called the Egyptian wheel was sometimes used. This consisted of a chain of buckets, and was used to draw water from deep wells. In 1600 an endless chain of thirty-six buckets driven by three horses was introduced into the Perthshire colliery of Culcross by Sir George Bruce of Cannock. This 'wonderful' colliery was visited by King James, involving the descent of a shaft. In 1602 the Hostmen of Newcastle were incorporated to regulate the coal trade, with an annual export to London of 190,000 tons using 200 hoys (boats) sailing between the Tyne and the Thames. No wonder that Tyneside was referred to as 'the Black Indies'. The bulk of the coal came from the twenty to twenty-five collieries on both sides of the Tyne west of Newcastle. People were now becoming more familiar with the use of coal, and King James made it known that he did not share Elizabeth's dislike for 'the foule smoke of sea-cole'.

In sinking, the 'adventurer' was guided in Elizabethan times 'by the judgement of those that are skillful in choosing the ground for that purpose'. Their skill depended largely upon the extent to which mines in the district had been already exploited; every new success in reaching coal added to the data available concerning the nature, the position, and the declivity of local seams. Even in the most extensively exploited districts, the miners sank many useless shafts. Prospecting charges were reduced somewhat at the beginning of James I's reign by the invention of boring rods, which made it possible to 'try' the strata without digging pits. Boring was the first process associated with deep mining, known from early in the seventeenth century, and was alluded to in Rovenzon's *Metallica* published in 1613 and according to Gray in *Chorographia* (1649). The first boring methods were so crude that trial shafts still had to be dug to obtain further information on the nature and thickness of the seam to determine the viability of establishing a colliery.

The mining expansion concentrated on deeper shafts over 15 fathoms, and drainage from the pits benefited from the introduction of German pumping techniques. Some of these deeper workings were in areas where surface coal had been exhausted. Deeper shafts required more capital investment, and this finance would have been lost if seams had been missed during boring and sinking. Therefore there was a need for precision in locating the coal seams. A group of exploratory sinkers (borers) employed by Nottingham Corporation were paid a somewhat exceptional 10d a day in 1594–5, compared to local miners at 6d to 8d. However, boring at Teddesley Common in Staffordshire was unsuccessful because the boring rods failed, breaking in a hole. Where coal was found by boreholes, deeper shafts known as 'Jacky Pits' were then sunk to depths of 90–120ft. William Watkins, for example, in 1600 sank such shafts to a depth of 94ft at Beaudesert Park in Gloucestershire; Shaft No.1 cost £14 16s 4d to sink.

A—AXLES.   B—WHEEL WHICH IS TURNED BY TREADING.   C—TOOTHED WHEEL.
D—DRUM MADE OF RUNDLES.   E—DRUM TO WHICH ARE FIXED IRON CLAMPS.
F—SECOND WHEEL.   G—BALLS.

A—TOOTHED DRUM WHICH IS ON THE UPRIGHT AXLE.   B—HORIZONTAL AXLE.   C—DRUM
WHICH IS MADE OF RUNDLES.   D—WHEEL NEAR IT.   E—DRUM MADE OF HUBS.
F—BRAKE.   G—OSCILLATING BEAM.   H—SHORT BEAM.   I—HOOK.

Left: Water pumping by German miners in the sixteenth century, *De Re Metallica* Book IV p.197.
(National Mining Museum) *Right:* Whim machine used by German miners in the sixteenth century, *De Re Metallica* Book IV p.166. (National Mining Museum)

In Cumberland, Sir Thomas Chaloner granted fifty-year leases to his tenants in 1560 which included the right to mine coals, and in 1586 he allowed St Bees School to take coal from his pits. Coal was extracted by means of a 'bearmouth', which was a tunnel within the seam opening to the surface at the outcrop. The coal was brought out through the bearmouths by carrying or dragging containers through the galleries or more often up shallow shafts using a hand-powered windlass or horse-powered gin. Flooding of mines from the sea could cause massive unemployment. Coal owner Christopher Lowther exported coal to Ireland from 1632, and this was probably from outcrops, however, forty-five years later it was recorded that the Mostyn Coal District of North Wales was supplying most of Ireland's coal from the River Dee.

The coal industry on the border between England and Scotland through the centuries has had to deal with continuing conflicts. In 1640 after the Newburn Battle, the Covenanters took possession of Newcastle; about 10,000 people employed in the coal trade fled. Despite the fact that Newcastle's walls were starting to decay, they were considered enough to protect the town's coal trade from Scottish raids. From early times, the miners involved in warfare were vital to the means of attacking fortresses. These miners (later called 'sappers' by the British Army in India) were also known as combat or assault engineers, and their engineering role included the clearing of minefields. The tunnelling up to the fortress walls was always dangerous work since it was strongly resisted by the besieged.

One of the first uses of 'the engine' was in making siege to castles. In the days of King David of Scotland, it was accepted that no one could ever succeed in conquering Newcastle without a battering engine (*sans engin*). In 1644 during the border conflicts at Newcastle, the besieged, with their countermines, had nearly succeeded in blowing up the chambers in which the Scots kept their mining powder. Under the direction of John Osborn, 'a false Scot', the colliers of Elswick and Benwell successfully sprang one mine near the White Friar Tower, and another at Sandgate. The batteries effected four great breaches; however, these were very steep and difficult to mount. Therefore the townsmen were able to inflict heavy losses on the Scots who were only armed with the iron spikes of their hand grenades. The colliers soon returned to the risks of their normal work.

Scottish mineralogist George Sinclair in *Hydrostaticks* (1672) raised doubts on mining methods at that time:

> To find Coal, where never any hath hitherto been discovered ... there were three wayes. First by sinking, which is most chargeable ... There was a second way invented to supply this defect, which is by boaring, with an instrument made of several Rods of Iron…

He also noted: 'I have known a Coal bored, with the Boarer … hath judged four foot in thickness, yet … hath not proven one.' His third method was by ranging or estimating the whereabouts of seams from their observed outcrop or dip. There was still some doubt in the mid-seventeenth century as to the merits of boring, and Sinclair and later Robert Plot, Professor of Chemistry (in his book, *Natural History of Staffordshire*, 1686) considered that when the coal laid deep, boring became scarcely less tedious and expensive than sinking trial shafts: 'the drawing of Rodes consuming so much time, in regard it must be frequently done.'

Mine owners were continuing to search for better means of ventilation. One method was the use of natural ventilation pits sunk in pairs; when not sufficient, Robert Plot described how draught was increased by:

> Sending down their lamp (fire bucket) into the shaft or by-pit next to that they intend to work, which makes a great draught of air from the bottom of the works, the smoke (gases) must necessarily come away and fresh air from above come down the other pit where the workmen went down, and the coal is drawn forth; it is however very costly.

An account of 1681 written of the collieries near the Mendip Hills, stated 'many men of late years have been there killed, many others maimed and burnt; some have been blown up at the works' mouth, the turn-beam which hangs over the shaft has been thrown off its frame by the force of it'. Often ventilation and drainage problems went hand in hand. As shafts went deeper, more thought had to be given to the task of draining the mines.

At Clifton coal mines, the famous civil engineer James Brindley used an immense waterwheel to clear the mines of water by means of the River Irwell. Chain pumps driven by waterwheels were employed to drain the principal collieries in the north of England, namely Heaton, Jesmond, and Ravensworth collieries on the Tyne, and Lumley on the Wear. In 1672 some of these collieries had attained the depth of 40 fathoms. The water, however, was not raised in one lift, but in several stages. At Ravensworth Colliery, water engines were used, thought at that time the most remarkable in the north of England. The total depth was divided in three stages with the water raised at each stage by a separate engine and pit. Three waterwheels were required, all driven by the same stream; one placed on high pillars, the second on the surface of the ground, and the third under the surface. The power to work the second and third stages was transmitted from waterwheels by means of vertical shafting and wheel-work placed in respective pits. At Lumley Colliery during the same period, there were two of these engines (one of three stories, and the other two) which served to drain all pits within a radius of two to three miles. In 1676, Lord Guildford owned coal mines at Lumley Park, Durham, considered the best in the North. His biographer noted: 'Coal lies under the stone, and they are twelve months in digging a pit. Damps or foul air kill insensibly: sinking another pit that air may not stagnate is an infallible remedy'.

The practice of coal mining and sinking up to the beginning of the eighteenth century (before steam power had properly developed) was set out in a treatise printed in 1708 entitled 'The Compleat Collier; or The Whole Art of Sinking, Getting, and Working Coal Mines &c., as is now used in the Northern Parts, especially about Sunderland and New-castle'. This treatise was in the form of a dialogue, firstly between a coal owner and a master sinker and secondly between the same owner and a viewer. The sinker (or 'sincker') advocated the preliminary operation of boring to determine the thickness of the coal bed, which he stated could be effected at a cost of 15s or 20s per fathom, whilst actual sinking might cost 50s or £3 per fathom. The sinkers' wages started at 12d or 14d per day. Only experienced sinkers should be used, for:

> … if he (the sinker) be altogether unacquainted with this sort of sinking labour, he may lose his life by styth, which is sort of bad, foul air or fume, exhaling out of some minerals, or partings of stone, and here an ignorant man is cheated of his life insensibly; as also he, by his ignorance, may be burnt to death by a surfeit, which is another sort of bad air but of a fiery nature like lightning, which blasts and tears all before it.

The author noted after reaching coal: 'My Business of a Sinker is at an end.'

Steam Power Fuelling the Industrial Revolution (1700–1900)

During the course of the eighteenth century almost every economic change increased the demand for coal for heating long before it was needed for power. Efficient pumping devices were necessary before deep and prolonged pit working could be ensured. Steam power was legitimately regarded as the 'all-powerful agent' in the mastery of nature. The application of steam was to impact on all aspects of industry. British aristocratic and land-owning entrepreneurs, unlike their European counterparts, owned the mineral rights under their land, and therefore they had a direct interest in new techniques which would increase their ability to exploit these resources.

The methods described in mining treatise 'The Compleat Collier' would have been defeated by seams below 60 fathoms, and therefore rumours of the invention of a stronger pump was very much welcomed by colliery prospectors. Gradually as mines had to be dug deeper to reach coal, the ingress of water into the shaft became the most demanding problem. Previously, drainage channels had to be dug but these were not always effective, and elaborate systems of water powered pumps were about to be developed. On the continent, French capitalists were also concerned with the expense of keeping the water back once the shafts were sunk rather than the costs associated with the initial sinking. An experienced observer of the Belgian coalfields described the miners' battle with underground water as the history in brief of coal-mining exploitation before the introduction of steam pumping engines.

The Newcastle owners were amongst the first to see the great advantages of steam power. In 1714 there were only four steam engines in existence, two of these being used in the Newcastle coal mines. The first Newcomen engine in the North East was used for pumping at Washington Fell Colliery, after being used to draw water from a mine

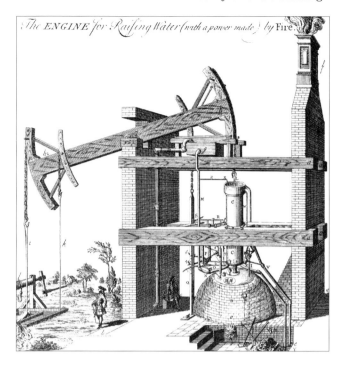

Newcomen pump, or
atmospheric engine, patented in
1705. (Science Photo)

near Wolverhampton three years earlier. Other coalfields were slower to adopt these
improvements in steam power. In Somerset the first 'fire-engine' was used in 1781 for
pumping water, and then for winding up coal; however, at many pits the use of horse-gins
was to continue throughout much of the eighteenth century.

In 1720 boring was used to find coal at Lord Molyneux's land at Sefton, in Lancashire.
If coal was found then the borers were allowed to use 'Boring Nagers and Ropes and
Utensils'. However, in the North East, master borers were acknowledged in 1760 as experts
in mining exploration; they knew the nature of the seams in an area twenty miles around
Newcastle to a depth of 600ft. Andrew Wake from Newcastle was recorded as boring for
the Ayrshire Coal Company in 1788–9, taking a year's work to get down to a depth of
378ft. As early as 1772, borings had been made through the overlay (overburden) and after
twenty years it had penetrated to depths of 505ft.

Daniel Defoe, during his *Tour through England and Wales* in 1724, noted the pleasantness
of the landscape around Lumley, but also that it was 'full of excellent veins of the best
Coal in the country for Lumley Coals are known for their Goodness, at London …' A less
complimentary comment made by the Duke of Cumberland, arriving in darkness before
the ravine at the bottom of which flows the River Tyne, who was heard to exclaim, 'For
God's sake, don't think of taking me down that coal-pit at this time of night'.

In the early 1700s, a Newcomen pump was used in a deep pumping shaft sunk at Howgill
Colliery in Cumberland. Due to local topography, this county was to develop deep-shaft
mining nearly a century before the Newcastle area. William Brownrigg, an eminent
scientist living in Whitehaven, was the first to investigate the explosive gas fire-damp. In
1741 he had married Mary Spedding who was related to viewers Carlisle Spedding and
his brother John. These viewers pioneered the use of explosives in sinking shafts and on
improvements in ventilation, also inventing a safety device called the Spedding Wheel or
Steel Mill. Candles used by coal miners continued to be a heavy expense suffered by
owners and contractors. Using just picks and shovels, gangs of specialist workers (sinkers)
excavated the pit, negotiating variable rates of pay to reflect depth and difficulty. In 1718

Carlisle Spedding started as manager of pits at Whitehaven, and he immediately introduced the building of a steam engine (or fire engine as it was then called) at a pit in 'The Gins'. Spedding arranged for a system of deep-shaft collieries with drainage in an integrated way for the whole area alongside development of under-sea mining. This proved successful, removing the need for the expensive and inefficient horse machines previously used for draining the mines. He carefully supervised the sinking of the Saltom Pit, Cumberland, in 1729. An early innovation by Spedding was the oval shape of the 10ft by 8ft shaft, divided down the middle to enable coal and water to be drawn simultaneously when in use. Spedding allegedly travelled to Newcastle to spy on the latest developments in coal mining, but despite being burned in an explosion during his visit, later he was able to use the North East know-how in Cumbria. Tragically, in 1755 he was killed in a mine explosion. Co-operation between coalfields began to improve when in 1812 Newcastle viewer John Buddle advised William Peile (viewer at Whitehaven and agent to Lord Lonsdale) on the installation of winding engines in the Cumberland pits.

Ensuring that the rate of flow of water was within the capacity of the available pumps was no more than a routine aspect of sinking. In 1750 at Houghton Colliery, sinkers reported: 'Raised a great feeder of water which Feader is more than the Engin can manage by about 2 or 3 hours in 24 hours.' From the seventeenth century, methods were developed in the North East coalfields to line mine shafts and thus exclude percolation from water-bearing strata into the shaft. These methods, known as tubbing, were first carried out using timber planks for the lining, resting on a timber curb installed in impermeable strata. Solid wood tubbing was found capable of resisting a pressure of 200–300 psi. The usual form of tubbing with timber was used at Hebburn, Jarrow and South Shields in the form of planks or solid cribbing. At the same time (1749) a ventilation furnace was located at the bottom of a shaft at Longbenton Colliery, and a few years later at North Biddick in Durham a surface-sited furnace ventilation system called an air tube was in operation. Thirty years later at Wallsend B Pit, an underground furnace was installed, considered superior to the surface type.

Historian Eric Hobsbawn noted that 'steam engines were the product of the mines'. However, the more efficient James Watt type of engine took some time to replace the 'atmospheric engines'. Four years after Watt's birth, an important improvement in steam engines in 1740 was the change from brass to cast-iron cylinders. Prominent among the builders of engines and promoting other improvements in mechanical engineering in Newcastle collieries was the eminent viewer, William Brown. From 1756 he proceeded to build engines for draining nearly 100 pits, including three in Scotland. Walker Pit, sunk in 1758, was drained by three engines. Five years later an engine of colossal proportions was employed at Walker which had recently been sunk to 100 fathoms, the greatest depth yet reached in the Newcastle district. This engine, brought from Coalbrookdale Foundry, was 74in in diameter and 10½ft in length. Four boilers supplied it with steam, and it was considered 'the most complete and noble piece of iron-work that had up to this time been produced'. In order to accommodate the pumping and winding, 'sinkings' or 'winnings' shaft diameters had to be increased. John Wilkinson, the ironmaster of Shropshire, was aware of the importance of finding a fuel for his blast furnaces to take over from coke and charcoal, since deforestation was becoming a major problem. To achieve his ambition, in 1771 he needed furnace builders, pit sinkers, engineers, furnacemen, bricklayers and many more. However, some of the impurities found in coal could have a devastating effect on the iron product. In 1772 he wrote to Birmingham manufacturer Matthew Boulton on his success in using coal in his furnace. The Newcomen engine which the mine owners were using for their own benefit was a crucial element in Wilkinson's need to obtain greater blasting power. The wealth of Staffordshire coalfields could now be exploited.

As Napoleon was overrunning Europe, George Stephenson and his father started work as engineman and fireman respectively at Wylam-on-Tyne using a pumping-engine erected by Robert Hawthorne. Up to this time, drainage had been provided by channels called adits, but by 1750 they were limited by the pumping methods available. Steam was to come to the rescue. James Watt obtained a patent in 1769 on his adaptation of the Newcomen steam engine, and then joined forces with Matthew Boulton to begin pump manufacture. In 1776 the first engines were used to power pumps and produced only reciprocating motion to move the pump rods at the bottom of the shaft. For the next five years Watt was busy installing these engines, mostly in Cornwall, to pump water out of mines. In iron-making, the first steam engine was being used for purposes other than pumping water. From 1785, for forty years a contest between water and steam carried on, with water undoubtedly cheaper where a good supply was available.

While the Newcomen had made a dramatic difference in drawing water from the mine at Walker, the location of the deepest and most important colliery, the method of removal of coal up the shaft had not changed. An attempt to address this shortcoming was made at Hartley Colliery in 1763 by way of a patented arrangement using a steam engine by Joseph Oxley. Five years later James Watt inspected this arrangement and considered it sluggish and irregular having no flywheel. The steam engine at this time was a single-acting machine, and was ill-adapted for producing an even rotative motion directly. Where water was readily available, Smeaton from Leeds, later known as the 'father of civil engineering', developed a system using a single waterwheel travelling in the same direction, and reversing the drum by means of gearing, but owing to its greater simplicity, the double-bucket wheel was universally adopted. Steam engines were sometimes used to provide a supply of water, as Smeaton reported in his 'Comparative estimate of drawing coals by horses, or by a coal-engine worked by water supplied by a fire-engine'. In 1777, he designed a 'water coal gin' at the Prosperous Pit (Longbenton Colliery) which was able to do the work of sixteen horses and four men. These engines continued to be in use for a further twenty years until the expiration of the patent for the application of the crank to the steam engine in 1794. The facility with which Watt's double-acting steam engine could be applied directly to the drum shaft caused it to rapidly supersede all other methods.

William Wilberforce, in 1779 during his journey to the Lake District, made numerous references to mining. He first recorded that at many coal mines near Masham in Yorkshire miners 'had a great genius for Husbandry'. He travelled on to Wensleydale where he observed that if:

> … the Limestone has a mixture of Lead in it, it is a sign of there being good mines beneath. They then (almost always from the top of the Hill) bore 20 or 30 or as low as 7 or 8 fathoms, & if they be incommoded by Water they drive a Level into the side of the hill to let it off. When driving or sinking, the Wages £1, £2 or £3 a fathom. …The Hole is in general 1½ yard high and 1 yard wide. The Master finds tools, the Men Powder for blasting & Candles… The employment is esteemed a very unwholesome one. An asthma is the complaint it generally brings on. They seldom live to 60. Never any damps in their mine.

Travelling on towards Lancaster, he noticed nearby coal carried in little carts drawn by one horse.

Sir James Lowther and his overseer Spedding were reluctant to show their workings to strangers, but Wilberforce was able to obtain the following information:

> The Coal is not found Vein but in a kind of stratum or layer extending far and wide. They leave pillars of coal to support the incumbent Earth, & when they can find little they thin

these, & when they quit the Mine almost entirely take them away, which is call'd robbing the mine. (A great number of fathoms of solid Rock, suppos'd so firm as not to be in any danger falling in, did, so necessary are these Pillars. I think it was near Newcastle). The Works under Ground are very extensive & are carrid a great Way under the Sea. Their Caution about shewing them is perhaps occasion'd by their fears of their being a fire. They in general find the Coal by boring to a great depth, perpendicularly. Sometimes it rises from below on an inclin'd Plane, so that you can go in with Ease. The Place where it opens in this Way is call'd the Beer or Bear Mouth. The Shafts are very long & the miners who are let down in Buckets are in danger for the smallest Stone falling on their Heads might do them a material Injury falling from such a Height. The Common Mark of Coal beneath is a dirty, blackish, crumbling Soil on a freestone, or some kind of stone. They are greatly subject to Damps in these Mines, especially the fire-damp, so that in some of them the Light they receive is from the Wheel full of flints turning against a Steel, as flame not sparks sets the damp on fire. The most dangerous & unwholesome time of visiting them, Physician told me, was in the Autumn or latter end of Summer. There is a Steam Engine. They meet sometimes with a strong demonstration of some deug or internal convulsion of the Earth – Coal, Stone, Earth, & all kinds of Soil strangely jumbled & thrown together all at once in a bed of Coal, which upon removing that is found as before. The Coal is raised out of the Mines by a Horse & put into Waggons which hold I forget the exact Quantity but I am sure 1/2 a ton. These have Wheels of cast Metal, & there are Roads made on purpose for them on the South Hill where it overlooks the Town, which have an inclination from the Mine to the Town, & have a piece of Wood on each side (sometimes two abreast) that exactly fits the Wheels so that the Carriages can in most places be push'd by men. These Roads terminate in a large Warehose from which there are six Great funnels which go at the bottom into Holds of the Ships & the Bottom of the Carts lets the Coals into it at the Top. The funnels are called Hurries. They have a contrivance to prevent the Waggons going too quick, a piece of iron, I think, pressing upon the Wheels. Lest there should not be ships ready receive the Coal, there is another Floor, beneath, of the same Warehouse, in which it is dropped in the same manner as in the Hurries & in which it lays till there is a demand for it …

In 1791 Arthur Young also observed that: 'all the activity and industry of this kingdom is fast concentrating where there are coalpits.' He should perhaps have mentioned that other rare materials in close proximity such as iron were also important. The combination of these materials promoted the development of transport in these industries and also affected the changes in methods used in sinking. Local well sinking to supply the ordinary water requirements of communities continued up to the introduction of piped water. Robert Southey on his travels through Birmingham recorded seeing a 'street of brick hovels, blackened with the smoke of coal fires which burn day and night in these dismal areas … the whole earth beneath us was on fire; some coal mines had taken fire many years ago and still continued to burn'. William Wordsworth's fire at Dove Cottage in Grasmere burnt sea-coal from the west Cumberland coast, whereas poorer families burnt peat.

At this time, James Hutton, founder of modern geology, was taking an interest in mineralogy, methods of coal mining and canal building, and perhaps laid down ideas of evolution which were to be of interest to Charles Darwin. Hutton was born in Edinburgh, and particularly in Scotland coal played an important part in the revolutionary improvements in agriculture. Coal was used to burn the limestone which then was used to fertilise the soil. The cost of keeping gin horses was as great as the wages of the gin keepers, but on the other hand for a small mining operation the cost of erecting a steam engine was often a very burdensome debt. The average cost of 'winning and sinking' a North East

'Unloading coal from corves at top of shaft at a Newcastle Coal Pit', lithograph by J. Christie (undated). (Stanley Library Durham)

Loading of coal using horse-wagons, pack animals and barrows during the early nineteenth century, painting 'A Pit Head' British School, c. 1820. (National Museums Liverpool)

colliery in 1830 was about £60,000 (compare this to twenty years later, when £24 million was being invested in the region's coal industry). This was mainly due to powerful owners wishing to gain access to the deeper coal seams. There was a large increase in manpower; from most mines having fewer than fifty miners, quickly the average grew to more than 200 miners. The dominance of the coal trade in the North East was highlighted in the Geordie pitman's song:

> I tell the truth you may depend:
> In Durham or Northumberland
> No trade in them could ever stand
> If it were not for the coal trade.

Ireland had lacked mineral resources, and its industrial development was severely limited by the restrictions imposed by the British Government. There was only small-scale coal mining at Arigna near Leitrim, worked as drift or open-cast, and at Castlecomer mines in Kilkenny where sinkers have worked from the 1790s. Ireland also had the disadvantage that its mines were not close to the sea. In 1809 the Arigna Colliery, employing 253 men in Roscommon, was estimated to have lost its backers £60,000 due to lack of management control. This problem was also observed in 1844 at Lord Wandesford's Leinster collieries, as 'the working was carried out without any rule by every small tenant, not only a great deal of valuable coal was lost ... but another large proportion stolen, or sold by the master colliers, who had the workings in their respective hands without accounting'.

Surveyor, geologist and drainer William 'Strata' Smith later to be known as Father of English Geology was already experienced in the Somerset coalfield. During his visit to Tyneside in 1794, he persuaded Colonel Braddyll, a major North East landowner, to undertake exploratory borings through the magnesium limestone. James Ryan who had been engaged as a mineral surveyor under the Grand Canal Company in Ireland, in 1804 introduced a method of boring allowing cores to be withdrawn for examination. For boring, as with the second process of shaft sinking, the most important changes came with the introduction of power-driven devices. Bore-rods were purchased at Newcastle upon Tyne in 1818 by Lord William Howard of Naworth Castle; and ten years later the same nobleman sent a borer to the Forest of Dean in search of coal. In Durham the main families carrying out boring were the Rawlings and Stotts, described as being 'the only respectable and professional borers in the north'. Usually, a boring team consisted of a master borer and four men, the works being undertaken by contract. Drilling or boring at depth for water was to gain from developments in general sinking. Herbert W. Hughes' *Text Book of Coal Mining*, published in 1892, mentioned that the boring equipment had not changed much during the previous sixty years.

In 1818, whilst the celebrated clown Grimaldi was performing in Newcastle, he was persuaded to see a coal mine. As soon as he went down the pit Grimaldi made it clear that he wanted to go back up the shaft without delay. Also Dickens described how he descended 200 or 300ft in a basket, and at the bottom of the shaft a piece of coal weighing about 3 tons fell close by. The future ruler of Russia, Grand Duke Nicholas, was also anxious to go down a coal mine and was therefore invited down Wallsend Pit by the viewer, John Buddle. However, on seeing smoke issuing from the shaft mouth, Nicholas decided not to go any further. Another visitor to the North East, accompanied by the viewer, John Elliott, described his sensations as he stepped from the edge of terra firma while swinging over a black depth of 1,000ft with nothing between the basket's bottom and the bottom of the shaft:

> Now we are ascending. People who go up the balloons affirm that they have no feeling of motion, but that the earth seems to be flying away from them, while they are sitting still and resting. Much the same may be said, in reverse, of descending a coal-shaft. You have no sense of descent; but the little round hole of light seems to be flying faster and faster over your head upward, as if it were going to the skies; and at length – in a couple of minutes, perhaps – the orifice of the shaft has apparently changed itself into a day-star,

Left: Sinkers repairing a drainage garland in the shaft lining, published in the magazine *The Graphic*, 28 September 1878. (British Library Newspapers) *Right:* Sinkers sinking a shaft in a coal mine, published in *The Graphic*, 28 September 1878. (British Library Newspapers)

which shines far, far above you in the firmament. The first occurrence that brings you to a consciousness of your rapid descent is, if it so happens, the up-coming of another load, and the passage of that load by you; including, as it does, if a load of live human stock, some one with a candle. And now, before you can have thought of it, you come close to the bottom. You have been close to four minutes in descending. Now you feel a full stop, which convinces you that you have previously been in rapid motion.

He had just descended a wonderful chimney of 1,600ft height. Unlike in a balloon where you are continually rising into pure air, in a coal shaft you sink into fouler and more fiery gloom, particularly since it was an upcast shaft with the furnace below ventilating the noxious gases.

In Ireland the cutting of peat or turf has been carried out right up to present times, mainly for domestic fuel. Before the start of coal mining during the 1820s, in Yorkshire peat cutting took place in a small way. Belgium, having considerable coal resources, was the second nation after Britain to industrialise. The British coal industry expanded rapidly during the 1840s and 1850s with the development of the railways. The exploitation of the east Durham coalfield went into full gear after the establishment of rail links in 1829. There was a great demand for sinkers, such as the colliery at Thornley, where the sinking commenced in 1841 and specialist shaft sinkers arrived from Cornwall and Germany.

The sinking of coal mines in north-east Durham revealed that there were large quantities of underground water in the magnesium limestone and, to exploit this, the Sunderland Water Company was established in 1846. The company enlisted the help of Thomas Hawksley, famous water engineer, who had pioneered piped water in Nottingham, the first such system in the world. His father had used a steam engine to grind corn at his mill to provide bread at reduced prices for the poor. Hawksley was involved in pumping stations being constructed at Fulwell, Cleadon, Ryhope and Dalton in Durham, as well as in Sweden and Germany. Many waterworks schemes were developed throughout Britain in the second half of the nineteenth century, and these drew on the sinking skills available, in particular those of boring. Durham master sinker William Mason worked on waterworks schemes in Lancashire. In 1856 piped water became available in Seaham for the first time; previously water had been collected in buckets!

In 1872, William Edward Nightingale granted a lease to the Stanton Iron Company to extract coal from near the Nottinghamshire/Derbyshire border close to the village of Pleasley. His famous daughter Florence is believed to have cut the first sod to commence the sinking in the following year. At this time due to high prices and rising demand, Durham coal was entering an unprecedented period of prosperity and shafts were being sunk all over Durham as fast as sinkers could sink them. An engineer estimated that sinking two shafts cost £1,800 with a further £2,200 for engines, gear, etc. At this time, Prussian forces were battling their way over the coalfields of Alsace and Lorraine and laying siege to Paris, so the coal owners of Durham were busily taking advantage of the rising price of coal stimulated by disruption of supplies from Europe. Colliers from Sunderland delivering coal up the Seine were seized and scuttled by Prussian troops. In the 1860s Prussia had been the leading industrial region of Germany, with great coal, iron and zinc reserves in Westphalia along with the coalfields of the Saar valley and Silesia. Britain and Prussia were both to gain from the exchange of mining engineers and sinkers. Many of the difficult sinkings on the Continent during the 1860s and 1870s were carried out by English contractors, but then British technology fell behind the new systems being developed by the Germans and French.

The introduction of grouting processes in sinking commenced with the publication of Beaudemoulin's last paper in 1851, and W.R. Kinipple carried out experiments in cement grouting. Thirteen years later P.W. Barlow obtained a patent for a tunnel shield, and English engineers pioneered shield-driven tunnels using grout behind the lining. In 1876 Thomas Hawksley used cement grout to seal fissures in the rock beneath an earth dam at the Tunstall reservoir in Weardale. Shortly afterwards, rock grouting was introduced to the coalfields of Northern France, and then there was a rapid development of injection processes, at first in mining and later in civil engineering such as dam foundations.

In 1899, Dawdon was sunk by methods gained by lessons learned during sinking at Seaham. They had the benefit of the Poetsch freezing method to overcome the water problem. A French sinking company was engaged to freeze the water and sand below the limestone, but due to unsatisfactory performance, a Belgium company took over the sinking operations. In 1902 the Theresa shaft sinking at 350ft had to stop to allow the sinkers in the Castlereagh shaft to catch up. During the following year with both shafts through the limestone, the freezing process was handed over to the German contractor. A series of boreholes, twenty-eight in all at each pit, was drilled to a depth of 484ft, tapering from 10in diameter at the surface to 6in at the bottom. The Germans also had to overcome the fact that the water level in the shaft varied tidally since the shafts were close to the sea, which slowed the cooling process. To counteract this, the bottoms of the shafts were plugged with 190 tons of concrete. However, in 1904 a sudden inrush of water killed a sinker; the works were abandoned until 1907. A German company was brought in to apply a more advanced freezing agent to keep back the water in sinking the 22ft diameter shaft down to 2,300ft.

In 1899 Portier used the injection technique to seal leaks in the timber lining of No.3 shaft at Courrieres, France. During the next five years eighty-eight shafts in France, Belgium and Germany were treated in a similar manner. Portier noticed that during the sealing, the flow in a neighbouring shaft was greatly reduced, and this led him to become the inventor of the 'cementation' process which was to have such immense importance in shaft sinking. After the initial application of cement injection to shaft sinking, mining engineers became interested in the possibility of grouting the sands and sandstones which often overlie coal measures, for they realised that such a process would be cheaper and quicker than the Poetsch freezing process. In 1904 cement grouting was used for the Compagnie de Bethune, involving the boring of grout holes around the shaft with the use of a pump for the first time. A Belgian mining engineer, Albert Francois of Francois Cementation Co. Ltd, was to further develop the cementation method which was to become the basis of modern shaft sinking.

In 1909 at Thorne in Yorkshire, Pease and Partners invited Francois to assist in sinking a coal mine shaft. A freezing technique was used to deal with the problems which was affecting both boring and sinking. Again a German firm was used – Tiefbau und Lalten industri Aktiengesellschaft. The attempt to sink the first shaft was successful, but was sabotaged by the German workers at the outbreak of the First World War in 1914. On resumption in 1919, the stabilisation work was changed from freezing to cementation. Following the successful sinking at Hatfield Colliery, the cementation system was used in the stabilisation of St Paul's Cathedral and in the South African goldfields.

The last phase of new coalfields in Durham ventured under the sea. A sinker at Blackhall in 1909 recalled that other sinkers at that colliery came from various places – some had stayed in the area after sinking nearby collieries, others had been involved with building of docks at Hartlepool and building of viaducts at neighbouring Crimdon Dene and Dene Holme. Irish immigrants and tin miners from Cornwall made up the workforce. Sinking of colliery shafts was to continue in the twentieth century, with only marginal improvement in safety as some later major disasters in Britain were to prove.

In 1952 sinkers at the Hawthorne Combine (Murton, Eppleton and Elemore collieries) were able to take advantage of the considerable development that had taken place in mining technology. The waters of the sand feeder were therefore negotiated by the freezing technique first used at Easington Colliery some fifty years before. By 1957 shafts were penetrating to depths of 1,500ft, lined with concrete to its full depth and divided into two halves, one side drawing coal from the upper horizon and on the other side its coals from the lower horizon. By 1971 Hawthorne was the largest coal drawing shaft in Europe, raising 51,694 tons of coal from the three mines.

Cornish Miners and Mineral Mining

It is important to include the story of the miners and sinkers working in all areas of Britain and Ireland, since their struggles have had significant impact on each others' lives and shaft sinking. This is of particular regard to Cornwall where minerals were to be found in abundance from early times, and which was probably the main reason for the first invasions of Britain. Cornwall was to become the 'Engine-house of the Industrial Revolution'.

Mining superstitions and folklore were part of Cornish tradition, and some of these have spread to other coal-mining areas. There has always been a close connection between Cornwall, Wales and Ireland. St Piran, probably of Irish origin, became the patron saint of Cornwall, arising from his popularity with Cornish tin miners. The missionaries from Ireland and Wales converted the Cornish people to Christianity. Before miners started their working shift in Ireland, it was customary to say a prayer below the holy cross or religious figure often located at the mine entrance, and a similar custom was adopted in Cornwall and on the Continent. Even the Methodist revival in the eighteenth century with its stern Puritanism, was unable to stamp out all of the ancient Celtic traditions.

Other than patronymic surnames, little of the early Cornish language has survived, however, some Cornish words have been adopted in other mining areas. For example, cauldron was the name given to the open-topped wagons for transporting coal, first pulled by horses, next by winches and then by locomotives. In Irish folklore it was where leprechauns keep their treasure. The caldera (Latin for cauldron) was a subsided pit left after volcanic action. In Somerset, the fossil root or fern on the roof of a coal seam was known as Caadron-arse, which gave a warning of a potential weakness, and derived its name from its appearance like the bottom of a cauldron. The word 'kibble', the metal tub which carried the sinkers up and down the shaft, and 'bratticing', the timber partitioning of the shaft, originated from the Cornish mines.

Cornish miners have long held superstition of demons known variously as the Knocker, Knacker, or Bucca, being the equivalent of the Irish leprechaun, English goblin or Scottish brownie. In the United States they were known as Tommyknockers. These small creatures were believed to wear miners' working clothes and they carried out random mischief such as stealing miners' unattended tools and food (more likely due to the mice or rats attracted to the miners' pasties!). Their names were derived from the apparent knocking on the mine walls that took place just before a 'cave-in' – the creaking of earth and timbers giving advance warning before giving way. For some miners, the Knockers were malevolent spirits endeavouring to cause a collapse. According to some Cornish folklore, the Knockers were the helpful spirits of miners who had died in earlier accidents in the many Cornish tin mines. These beliefs were also adopted by miners who had emigrated to the Pennsylvanian coalfields, and much later to South African mines.

Cornwall held the richest natural resources in Europe, and therefore attracted adventurers from far and wide. Centuries before the Christian era, Phoenician traders came to Britain in search of tin, a particular product of the Land's End district. It was in connection with traffic in this metal that Britain made its first appearance on the page of history, and received from ancient writers the name of Cassiterides, or Tin Islands. The principal known metals at this time were gold, silver, copper, tin, iron, lead and mercury. The Romans also came looking for metal, separating tin from gravel before mixing it with copper to produce armour, helmets, shields, jewellery and goblets. Apparently the Romans worked in the North Molton mines in Devon, and during the reign of King John, these royal mines were known to be also rich in gold, silver and copper. The lead of Derbyshire was undoubtedly worked by the Romans as well as iron mining and smelting in the Forest of Dean. In those days, to dig in mines was considered the most severe punishment for criminals. The persecuted Christians who escaped death were made to linger out a miserable existence in 'those dreadful places', often working in water so stagnant that they frequently died.

The mining of minerals in Cornwall and in other counties have contributed to the general development of sinking in Britain. The skills of shaft sinking were learned in Cornwall and Devon before other counties due to the geological nature of the ground where veins of ore were often found in vertical or steeply inclined seams or faults. These skills were adopted later in areas such as the Pennines where most lead seams were also near vertical. In coal districts, the coal was laid down mostly in near horizontal seams allowing low-level exploitation without initially the need for deep shafts. In the Alston Moor and Allendale areas of Northumberland, lead ore has been worked since Roman times. The earliest method of working the ore appears to have been by sinking shafts as in the case of coal mines. Due to the irregular nature of the mining work, men would combine this work with farming in order to eke out some form of existence. This type of mining was the oldest industry in the Peak District, probably beginning in pre-Roman times by open-cast extraction. The Romans would have used lead products such as pipes, water tanks, roofing, lining to fonts and coffins. Deaths in these mines were mainly caused by rock falls and 'drownings', although a burial register recorded in 1669 the death of two miners who had suffocated. The lead mines were given certain royal concessions, and were mostly run on a private basis until the start of the Industrial Revolution. The mining was carried out initially in a primitive manner with simple tools such as picks, involving sinking short shafts, and then picking away at the ore in confined spaces. This progressed to heavier hand tools, followed by use of plug and feathers and fires to blow off the rock.

In the twelfth century, the Charter of Stanneries was declared. This outlined the lawful means of claiming tin from the land:

> Other than payment of a toll to the lord of the manor, for the first time in history, the miner was his own master, independent, not a serf or hired labourer. The prospective tinner would make a claim on a plot and would indicate its location by 'bounding' (sinking a pit and lining it with timber). As early as the year 1450, tin excavation took place by means of surface workings that were open to the sky. At first it was carried out using rope and primitive ladders and hand tools. The open pits were called coffins or gobbins ranging from a few feet up to 50ft deep. Coping with incoming water was a rush against time using primitive pumps, and waste material being removed by windlass. Tin lodes often outcropped on the cliff faces, and working conditions were appalling, with high winds and men balancing on jagged rocks. A century later adits were used to drain away water, having tried teams of horses and every possible system of mechanised waterwheel. Once the tin had been washed it was carried on packhorse or mule to the nearest blowing house for smelting. After the weighing, taxing and stamping with the Duchy arms, there were great occasions for festivity. Several days would be spent enjoying the profits. In several of the town's inns, or beside the stalls of market traders, hurling, drinking beer or cider, wrestling and cock fighting were the pastimes enjoyed by tinners.

With the departure of the Romans, the need for lead and mining skills went into decline. In medieval times only relatively small quantities were used on the building of churches and large houses for the roofs and drainage. Increased output from the mines was made possible by new techniques developed in Germany and brought to England by Dutchmen during the reign of Elizabeth I. A great expansion of mining of all sorts – lead, copper, tin, iron and coal – began with German miners opening up copper exploration in various parts of the Lake District. Daniel Hechstetter, a German mining engineer, and his sons were able to prospect in Cumbria in 1564, finding copper veins exposed in Buttermere, Borrowdale, Coniston and Caldwell. Vast quantities of lead were also found. The numbers of German miners were small compared with the large importation of foreign artisans in other industries such as glass and salt making, although Henry VIII did appoint Joachim Hechstetter and 1,000 men in his command to work in silver mines at Combe Martin.

Slate was extracted from Roman times, particularly in North Wales. Investigation has shown that a ship of the sixteenth century was found to be carrying finished slates. The men worked slates in partnerships of four to eight known as 'bargain gangs'. A 'bargain letter' confirmed agreement on the working of certain areas of rock between men and management. Letting day took place on the first Monday of each month. In Cardiganshire in 1604, it was recorded that the county was rich in lead and silver mines but the mines were so drowned in water, it was difficult to get at the metal. From the eighteenth century, like coal mining, the extraction of lead involved the use of pumping and winding engines, and to drain the mines, ambitious drainage soughs were built. In North Wales, these artificial watercourses, dug as open channels and supplying water to drive a waterwheel, were known as 'leats'. As mines became deeper, the cost of drainage became uneconomic and the industry went into decline. Thus, in the 1860s, lead miners from Blanchland, near Shildon in Durham, emigrated hoping to find a better life in Australia. An exception was Millclose Mine in Derbyshire, where the Watts Shaft reopened in 1859, and expanded to become Britain's largest ever lead mine.

A major advantage in the metal mines as compared with collieries was generally the absence of explosive gases. In Cornwall the lack of coal resources meant that fuel was always at a premium. In the reign of William III the traveller Miss Celia Fiennes on a riding tour of the South West found her supper in Penzance 'boiling on a fire always supplied with a bush of furze, and that to be the only fuel to dress a joint of meat and broth'. The Cornish forests had disappeared and the French privateers in time of war prevented the delivery of Welsh coal to the south Cornish ports. On the surface it was difficult to imagine that mining was being carried out below ground. Some of the Cornish mines were on a small scale with the works blending into the landscape as at the Carclase tin mine. Miss Fiennes also noted that men were even employed on Lord's Day to keep mines from flooding.

The use of explosives in sinking, rather than coal extraction, is thought to have been first used in the Cornish tin mines in 1689 although there is some evidence of its use in Staffordshire copper mines sixty years before. Black powder introduced by German miners was used in the tin mines in 1670. There is reference to the use of gunpowder for shaft sinking at a Somerset colliery in 1719, but normally gunpowder was used more often for deeper mining. The holes were charged and sealed with clay stemming. The miners pushed a hollow quill or straw through the stemming and filled it with finely ground gunpowder. Then a candle was fixed under the protruding end of the straw in a way that would allow it to burn for a short time before the fuse ignited. Premature blasts and misfires were all too frequent. Gunpowder was a low-power explosive and it could only be used to break off large lumps of rock near the face. At this time waterwheels provided power for crushing rock after blasting.

After Abraham Darby's successful use of coke for smelting in 1708–9, undertakings which mined both iron and coal often established their own furnaces. In 1763, a Welsh company, Dowlais Iron Co., developed on the basis of ironstone existing alongside coal reserves which were plentiful; coal and iron complimented each others profitability. In the 'heads of the valleys' ironworks in Monmouthshire, 'master miners' were appointed to raise both iron ore and coal. Ironstone had been worked in Northamptonshire from early times, but it was the Great Exhibition in 1851 which was to give the impetus to mass exploitation of ironstone in many areas of Britain. In the 1870s, the Staveley Coal & Iron Company was a powerful company based in north Derbyshire. Ironstone mining in the Cleveland area was to become the major industry in the early 1800s. Its extraction was relatively easy since it outcropped in many areas, and the deeper deposits were accessed by means of drifts and shafts. Ventilation and winding was achieved by means of fans powered by steam and later by electricity. Initially the ore was shipped to Newcastle where it was processed into pig-iron, and then transported back to Middlesborough where Bolcklow and Vaughan had opened their ironworks in 1840. There were also ironstone developments in Lanarkshire; George Emmerson was working there in 1851 as a young agent before he became master sinker in Durham.

In 1698, Captain Thomas Savery, manager of a Devonshire tin mine, took out a patent for 'raising water with the power of fire', and he described his 'engine for raising water by fire' in mines by means of a furnace and chimney as 'The Miner's Friend'. He was shortly joined by Cornishman Thomas Newcomen with his 'atmospheric engine' which was promoted as capable of discharging as much water as using fifty horses. He extolled its benefits as freeing mines from both water and from 'damps' or noxious gases. He was assisted in making further developments by Joseph Hornblower from Staffordshire. However, Newcomen's engine, with pump rods and a piston attached to either end of a beam engine, was not a great success since it was limited in the height it could raise water. Around this time the Cornish engine was described as, 'a filthy jumble of a thing, in which lots of spun yarn, pieces of rope, leather etc ... steam could be seen flying in all directions, and the arch-head chains could be heard at a distance of a mile or more.' These pumps were costly in terms of fuel consumption, but as demonstrated at Barnsley, they were able to raise 3.5 million litres of water every day and made it possible to mine annually 4 million tonnes of coal. This was a major improvement which meant new pits could be sunk and old ones reopened. The first Newcomen atmospheric pumping engine was erected at Coneygre Colliery in Dudley, Worcestershire, in 1712, and was to be the most powerful means of draining colliery workings for the next sixty years.

Portrait of Richard Trevithick, painted by John Linnell, 1816. (Cornish Studies Library)

The mining of minerals and coal required considerable physical strength, and therefore it was essential for miners to have plenty of nourishing food. In some counties like Leicestershire, miners enjoyed better health than agricultural workers despite their working conditions. Warwickshire miners had the reputation of being a tall and athletic race of men, like 'fighting cocks'. However, the colliers of south Gloucestershire were known as 'the terror of the surrounding neighbourhoods and for gross ignorance, rudeness and irreligion were almost without parallels in any Christian community'. Gradually, due to the efforts of the Dissenters, the miners of the Forest of Dean had replaced trespassing, outrages and savage amusements for regular attendance at chapel. In Cornwall and South Wales the miners' Celtic origin was blamed for their religious hysteria derived from their initial superstitions. In Ireland, Methodism never established a strong following.

As well as the famous inventor Richard Trevithick, many other Cornish men were involved in major developments in mining. Trevithick, jointly with his cousin Andrew Vivian, secured a patent on a steam carriage in 1802. Two years later he demonstrated his locomotive engine on the Penydarren Ironworks waggonway at Merthyr Tydfil followed by his visit to Newcastle upon Tyne with details of his locomotive. One year later he invented his light high-pressure boiler for portable purposes, in which to 'expose a large surface to the fire'. Sir Henry Beighton at Newcastle made further changes to the Newcomen engine following the improvisations by a boy named Humphrey Potter. Then Woolf, a Cornish engineer, patented an advanced boiler based on Boulton's work on the Wheal Busy engine. In 1831 William Bickford invented the safety fuse used in blasting, which was to save countless miners from an awful death.

But in Cornwall these achievements were extraordinarily one-sided. They made it possible to discover and bring up ore from the deepest levels; but for all their winding machines and pumps and their maze-like ventilation systems, they invented nothing to lighten the miners' workload. When it came to transport, the crooked shafts and slanting galleries seem to have defeated them. Men went up and down the mines by ladders and their work itself was untouched by machines. Although the Cornish sinkers were advanced in many ways, the primitive 'man engine' (a rudimentary and perilous version of the 'cage')

Idealistic picture of a Cornish miner about to start his working shift, painting 'From Under the Sea' by James Clarke Hook, 1864. (Manchester Picture Gallery)

was never generally installed. In 1846 men were rescued from the East Wheal Rose Mine near Penzance, by clinging to the kibbles and chains 'like strings of onions'.

The celebrated mines of Cerro de Pasco in Peru were to benefit from Trevithick's pumping engines in 1817. Shortly afterwards he personally visited these mines to sort out some teething problems, but a civil war intervened with machinery thrown down the shaft by the rebels. In South America the Cornish miners were known colloquially as 'Cousin Jacks', hard-rock miners *par excellence*. Nine years later, Robert Stephenson, acting consulting engineer in Colombia, was confronted by an insolent 'captain' of the miners who 'displayed a very hostile and insubordinate spirit', and who quarrelled and fought with his men. The captain and his gang, being Cornishmen, told Robert to his face that because he was a north country man, and not brought up in Cornwall, it was impossible that he should know anything of mining!

Swansea's copper industry relied on the import of ore from Wicklow in Ireland. After the end of the Napoleon wars, Cornish skills were in demand by Ireland and Wales. In 1821 captain Mark Reed arrived at the Dooneen mine in west Cork, and the following year, captains John Reed and Richard Martin recommended that a steam engine be purchased. Before Martin travelled, he obtained some training in copper ore assaying from Mr Jenkins, assay master in Redruth. Although there were improvements in Irish mines during the next twelve years, when Captain Samuel arrived from Cornwall, he did not find the workings to his liking, and he shortly emigrated to America where he worked at the galena (lead) mines in Illinois. In 1841 the commissioners, when reporting on the Tipperary collieries, stated: 'I am told the Irish are not clever at sinking shafts but are pretty good miners, so long as they have some experienced Cornishmen working with them.'

The general public was beginning to take interest in the mysterious world of the miner. In 1829 at the Theatre Royal in Worcester, a new melodrama was performed called *The Cornish Miners, or The Maniac's Den*. Was the playwright suggesting that miners were insane to carry out such work? The following extracts were included in the programme:

Act I, scene 3: The Shaft, or Entrance to the Miner
scene 4: Interior of a Tin Mine, Miners at work,
implements scattered about and mine
basket constantly ascending and
descending the Shaft
inc 'We merry miners few cares know'
Chorus
The stoppage of the Steam Engine.
The awful Breaking-in of the Mine, and
appearance of the **Maniac**

Act II, scene 1 The Shaft
scene 6 Gallery of the mine. Miners discovered
in despair. The rocks burst with terrific
explosion and the miners saved.

Last inc Finale Exterior of Mine
Scene 'Success to our Cornish Miners'

However, the Children's Employment Commission report of 1842 showed that Cornish mine workers suffered from the serious diseases of the lungs, hearing and digestive organs. Dr Barham drew attention to dangers of overexertions by miners due in particular to having to climb up and down long ladders in mine shafts. Underground workers were found to be inferior in physique to the female labourers who worked on the surface. Many of the men died young of galloping consumption. The condition of lead mines of Alston Moor on the borders of Northumberland and Durham were very similar to those in Cornwall. In both areas the Commission report included miners' complaints of lack of

Miners drilling into the rock at King Edward Mine, Camborne, *c.*1900. (Cornish Studies Library)

Miners working in a sloping lode supported by props at King Edward Mine, Camborne, *c.*1900. (Cornish Studies Library)

Miners unloading excavated material from a kibble into a tram using a mechanical winch at Cook's Kitchen, Redruth, c.1900. (Cornish Studies Library)

oxygen, excessive dust, and fumes of gunpowder. Engels' suggestion that miners in Alston did not start work until the age of nineteen was unlikely to be true. Matthias Dunn, a viewer from Newcastle, considered the methods of ventilation practised in Staffordshire and Wales to be inferior to the North East, and that in these other districts the boys were worked harder, with lower wages and longer hours. The 1851 census revealed that boys as young as nine were working as miners, and a 'mining girl' of eleven years was recorded.

However, from the point of view of technology and capitalist finance, in the 1840s and 1850s the Cornish tin and copper mines were more advanced than many of the collieries in the rest of Britain. The Cornish miners travelled and worked in 'pares' and 'these [miners] when the lode is wide, will sometimes all work together, but generally they divide themselves into parties, relieving each other at the expiration of eight hours.' The Cornish miner did not necessarily have to find his own working materials or tools – they might be loaned to him by the company and charged on his month's account. The 'tributers' were paid by the fathom rather than in proportion to the tin or copper raised. They bid for the pitch and also had the right to sink shafts and provide their own means of access and transport. Their earnings were more regular, but they had very little scope for speculative gains. The sinkers, known as 'tutworkers', were also responsible for driving forward the levels; they also rented their pitches, and had to bid for it on the monthly setting day as each working period started. Both tutworkers or tributers had their own choice of those working in their gangs. Father and son, or brothers, worked together on a pitch. Enginemen, binders, pitmen and head smiths were paid the most overall. Binders oversaw the maintenance of all timbers, and the pitman the maintenance of the shafts and ladders. Census records failed to identify the sinkers, but shaftmen and safety fuse manufacture workers were recorded.

In Cornwall, with virtually no owner or agent supervision, only 'captains' had a wider management role. They had an in-depth practical knowledge of geology to allow them to survey for further exploitation. A purser was appointed to carry out bookkeeping and to

Above: Sinkers with pump pipework at one side of the roughly excavated shaft bottom at Cook's Kitchen, Redruth, *c.*1900. (Cornish Studies Library)

Left: Sinkers drilling into rock and a young miner using a large hammer to force in a wedge at East Pool Mine, Camborne, *c.*1900. (Cornish Studies Library)

Croust time, with a large group of miners taking a break adjacent to a tramway at East Pool Mine, Camborne, *c*.1900. (Cornish Studies Library)

pay the tutworkers on regular wages and the tributers who were self-employed. Due to the variation of the ground, 'the lode is cut rich today and tomorrow poor'. This made the Cornish miners' life very vulnerable. For some time before the setting the mine captain or agent would go through the mine and decide on the work to be carried out during the next period. He would determine whether new shafts needed to be sunk or levels driven to develop the mine and he would judge how many pitches or divisions of the lodes were to be worked or 'stoped' in order to produce the ore.

In 1844 Charles Thomas was appointed manager or 'captain' of Dolcoath Mine in Camborne. He had started down the mine at the age of twelve, and only nine years later he was made underground agent. Long having been an advocate of deep mining, and despite opposition from shareholders, he persuaded a group of miners to work at the lowest levels of the mine. These tributers were paid solely on the basis of the ore recovered, so this activity was undertaken with no threat to the shareholders. Good deposits of ore were quickly discovered and this helped Thomas raise capital needed to deepen the mine further. In 1853 the first dividend was paid to shareholders and Dolcoath went on to become the most productive tin mine in Cornwall. With greatly improved pumping equipment, Cornwall began to develop deep-shaft mines. These consisted of a maze of tunnels and shafts interconnecting at different levels. Tunnels were worked along lodes which were rich in tin ore. Steps and shafts provided access at each level, often with a series of lodes running parallel with each other. Later shaft diameters became larger to accommodate their multi-purpose usage. More shafts were sunk vertically and some were sunk to depths up to 3,000ft with tunnels extending several miles under the seabed. There was a steady increase in trade to and from Wales, with sailing vessels taking ore to the smelters and returning with coal needed to fuel the pumping engines.

Cornish mines were literally 'adventures' – with an element of gambling speculation which meant that the company was put at risk when yields proved disappointingly low.

The tin and copper were thinly disseminated in very hard rock, and progress in boring was slow. Historian Joseph Y. Watson, author of *Compendium of British Mining*, wrote that 'one, two or three feet in a week, or a few inches daily, is often the whole amount of the united operations of twenty or thirty men'. However, the father of William Morris was willing to take that risk and he invested in a copper mine in Devon in 1845. A friend of zoologist T.H. Huxley, the scientist J.B.S. Haldane, had other interests and travelled to Cornwall to study hookworm in miners. He also carried out research to understand how carbon monoxide leaks killed miners, but unfortunately in the process of carefully taking and measuring his own blood samples, methodically poisoned himself.

The Poor Law of 1834 and assisted passages given by major steamship companies ensured Cornwall's loss and the world's gain. Emigration in small groups was actively supported by miners' unions. The Australian Agricultural Company sponsored many miners, and most miners were able to afford a single fare of £16 if they had been thrifty. When gold was discovered in 1851, coal had been mined at Newcastle (New South Wales) from 1801 and copper from 1843 in South Australia. This latter location was populated predominantly by the Cornish, German and Irish. The main equipment used in these mines was imported from Cornwall. For alluvial gold mining, the technology used was derived from both Cornwall and California. For example a wide range of hand tools were used including the English pick, the American pick and a number of single-pointed Cornish picks. There were a similar range of shovels. When shaft sinking was carried out, the single-pointed pick was favoured because of its lightness and ease of use in a confined space. The mining of quartz, that is stoping, was also executed using hand tools even into the twentieth century. The first rock drill used in Victoria was Low's patent rock boring machine which performed well at the Paris Exhibition. The drilling and blasting expertise of the Cornish miners was much respected and they were considered as imbued with great powers, by some who had never been down a mine. Cornish influence by 1855 was nearly total with engineers such as William Hosking and John Phillips designing mining plant specifically for Australia.

In the 1850s an influx of Cornish miners were attracted to the copper mines of South Africa. Anthony Trollope exploited his knowledge of Cornwall and its tin mining in his novel *The Three Clerks* in which he made it clear that he had himself had been down a tin mine. Trollope on a visit to South Africa in 1877 was to describe the activity at the bottom of a great pit as though seen from Dante's Inferno. This pit was 9 acres in extent and 230ft in depth, overhung by engines and shafts, pulleys and buckets for excavating the earth. As late as 1881, gold mining was still being carried out at the Clogan Gold Mine in Merionethshire. The Big Pit, incorporating Coity and Kearsley collieries, was so called since it was the largest shaft in Wales at that time, accommodating two tramways. Generally the mining of gold involved only shallow excavation, unlike in South Africa where shafts were to be sunk to nearly 3,000ft, taking seven years to complete. Cornish miners, or miners of Cornish descent, formed the overwhelming proportion of the skilled labour force on the Witwatersrand gold fields in the first twenty years of mineral exploitation. In 1902–3 the Miners' Phthisis Commission estimated that over 90 per cent of all white miners were of foreign origin.

Even today, the Cornish landscape still has countless ruins of redundant stone and red-brick engine house and mine buildings, providing a reminder of the once thriving mining industry. In comparison, the pit-heaps which once marked the North East, have been removed and landscaped, and beaches cleaned, so it is difficult to imagine how coal mining was once such a dominant industry. Into the twentieth century the health of mineral mining was to decline more rapidly than the coal industry.

Sinkers and Navvies on Canals and Railways

The 'navigation' or straightening and dredging of the rivers was the first major step in improving the transport of coal. Canals then provided the new watercourses for transportation. As steam power allowed mining to increase its scope in reaching the lower seams, steam locomotives then took over as the main means of transport for mined materials. Unlike coal mines part regulated since 1842, the sinking works on canal and railway construction had no such safeguards on working conditions.

The Duke of Bridgewater had the foresight to engage civil engineer James Brindley in 1761 to design a canal to serve his colliery at Worsley near Manchester. A few years earlier the Sankey Canal near Warrington, the first wholly man-made canal to be built, was used for the transport of coal. Even earlier, it was the opening of the Douglas Navigation near Wigan in 1742 and the subsequent joining of the Liverpool Canal at Dean Locks that caused the coal rush, bringing Liverpool and Bradford speculators into the area. Of these, the Blundells developed the extensive coalfield at Orrell. Mining at Hyde Colliery commenced in 1794, as soon as the Lower Peak Forest Canal was constructed. Similarly, the completion of a canal and waggonways near Sheffield in 1819 prompted the sinking of bell pits at Tinsley Colliery later that year.

The tunnelling techniques used for these canal works were 'borrowed from the mining industry'. Navvies specialising in building tunnels were known as 'tunnel tigers'. The first tunnels or horizontal shafts were simply openings through the hills, and when in use to pass through the tunnel the boatman had either to 'shaft' the boats or 'leg' them. Shafting, known as 'legging' in the North, was achieved by pushing a pole against the roof of the tunnel. This method was used at Neston Colliery in Cheshire which started its works in 1759. This colliery was located on the edge of the Dee estuary and its galleries were driven below the River Dee. These acted as miniature canals along which travelled 'starvationer' coal boats (named after their ribs), and the coal was then lifted up vertical shafts to be loaded into ships for export to Ireland.

In tunnelling, the worst accidents took place with falls of earth and rock, collapse of props and scaffolding, and during travel up and down the shafts. Construction of reservoirs to feed the canals was an important part of the works, and required a high level of civil engineering expertise. When the Diggle reservoir failed on the Huddersfield Canal in 1810, six lives were lost. Alas, the canal companies were reluctant to pay compensation. The Huddersfield Canal Company in charge of one of the largest tunnelling projects kept no records of tunnel accidents. As in mining, many disputes took place involving the managing engineers and the manual workers. A major reason for these disputes were the 'tommy shops' owned by the employers, which the workmen were required to use at exorbitant prices.

Towards the end of the eighteenth century Britain was gripped by 'Canal Mania'. This surge in the development of canals allowed many more mines to be economically viable due to the reduced cost of export of extracted materials. Following the second period of major canal construction from 1805 to 1830, railways took over as the main form of industrial transport. A new breed of labourer was born to provide the heavy physical input these massive projects needed. The railway labourers became known as 'navvies', having first worked on the navigations. After *The Times* coined the word 'navvy', this name started to be used by the general public to describe any labourer involved in heavy construction work. Although most of the navvies were sourced locally in England and Scotland, an increasing number came from Ireland, especially as famines forced Irish men to seek work overseas. On the Duke of Bridgewater's canal, the navvy was then an untrained labourer, but once he gained the skills required for these works, he was a much sought-after commodity by contractors. Civil engineers Thomas Telford and John Rennie were often required to employ local labour, but for vital canal works they ensured that experienced navvies were used. Labourers with similar experience such as the sea-wall builders of Lincolnshire were also recruited as 'navigators'. In addition, employed on the canals were men with specialist trades such as carpenters, masons and blasters who were paid more than the ordinary cutters or excavators. Thomas Telford and William Jessop supervising works on the Caledonian Canal explained why navvies could demand high wages:

> As canal work is very laborious, they must give such Wages ... as will be the means of procuring and calling forth the utmost exertions of able Workmen.

James Watt was employed as a surveyor for the cutting of one of the new canals in the Birmingham area. He then went into partnership with Matthew Boulton to produce steam engines which were to be used so widely in the coal industry. To superintend the day-to-day site works, a resident engineer was appointed as described in 1793 by Archibald Millar, Resident Engineer on the Lancaster Canal, as follows:

> A person capable of conducting the business of a Canal through, viz., that he is a good Engineer, can carry an Accurate Level, and has a perfect knowledge of Cutting, Banking, etc, and also that he is a compleat Mason.

On many of these canal and railway works, horses were used on the barrow runs to remove excavated material, and circular horse-gins were used as winches pulling waste up the tunnel's shafts. The lowest paid were the boys employed to lead the gin horses and to work on the barrow runs. Later, on the large schemes like the Manchester Ship Canal, completed in 1894, steam-powered excavators were used, which reduced some of the risks attached to manual and horse-powered working. These new machines were capable of handling large quantities of material. Navvies worked hard, and were well paid compared to labourers in other industries. However, at times they considered that they were not properly treated, and quickly showed the strength of their feelings. The Sampford Peverell Riot of 1811 in mid-Devon developed from a dispute between a contractor and his hired navvies. Anger at the token method of payment also led to rioting and violence in the small community of men working on the Grand Western Canal. One newspaper described the rioters as 'a savage ungovernable banditti'.

Both the canals and railways involved the construction of tunnels with the need for access shafts during construction, as well as shafts for permanent drainage and ventilation. Exploratory boring was also carried out on these contracts. During this period of extensive canal and railway works, the following projects illustrate where shafts were sunk and the disputes which took place on many of the projects.

Following trial borings, fifteen shafts were sunk during the construction of the Harecastle Tunnel during canal works just north of Stoke-on-Trent between 1770 and 1777 under the supervision of civil engineer James Brindley. Being one of the earliest canal tunnels, the engineering was very primitive, basically a rough hole through the hill. The shafts were used for removal of waste material and workers' access by means of horse-gins. The rock was blasted out by gunpowder, but the main part of the excavation was carried out by picks and shovels. Later rails were laid in the tunnel to facilitate waste removal. The construction site was subject to flooding, and steam engines were used to operate the pumps. Despite these problems, Brindley was able to assure his client, Josiah Wedgwood, that the project would be completed. Stoves were installed at the bottom of the upcast pipes to overcome the problem of ventilation. The contractor received £7 per yard, and the labourers £5 per yard, having to supply their own candles and gunpowder. Many canal works were delayed due to resistance by land owners, which was to be the case for the later railway schemes.

The Strood–Higham Tunnel was proposed in 1778 for use by military vessels between the Chatham and Woolwich dockyards, avoiding the need for a forty-six-mile round trip. A contemporary account gave the benefits of this project:

> The navigation of barges and other vessels on the canal would greatly facilitate and render less expensive the carriage of all kinds of wares, goods and articles, and would materially improve the agriculture of the circumjacent country, and would render unnecessary a long and circuitous and sometimes dangerous navigation on the open sea, and would otherwise be a great private and public advantage.

This was thought the first tunnel to be aligned with a scientific instrument. Shafts were sunk and two heavy plumb bobs used to obtain an exact alignment. On this project there were a number of fatalities including shaft deaths:

1820 – worker, forty-one, fell down shaft.
1821 – worker, twenty-nine, killed by fall of chalk.
1822 – three workers killed by fall of chalk.
1823 – worker killed by becoming entangled in rope going down shaft.

In 1780 on the Grand Junction Canal, wet conditions dictated the use of a through-heading which would act as a drain to the workings. The Revd Shaw visited the Sapperton Canal Works on the Thames & Severn Canal in 1788. Although he did not enter the tunnel, he had visited mines with similar conditions, which he described:

> Thus far in the mountain, with the aid of lights, 'tis easy enough to acces; but such a horrid gloom, such rattling of wagons, noise of workmen boring the rocks under your feet, such explosions in blasting, and such a dreadful gulph to descend, present a scene of terror that few people, who are not versed in mining, care to pass through ... On the passage down, the constant blasting of rocks, louder than the loudest thunder, seems to roll and shake the whole body of the mountain ... the glimmering light of candles, and suffocating smell of sulphur and gunpowder, all conspire to increase surprise and heighten apprehension ... at the same time figure to yourself the sooty complexions of miners, their labour, and mode of living, and you may truly fancy yourself in another world.

One of William Jessop's projects in 1794, the Butterley Tunnel near Derbyshire, the longest canal tunnel in the world, required thirty-three shafts with de-watering by means of a Woodhouse steam engine. Also involved was the provision of a service shaft which fed the canal from a local reservoir. The Lapal Tunnel near Dudley in the West Midlands was completed in 1798 after seven years, and construction required the sinking of thirty shafts. Initially two shafts were sunk along the line, and then small horizontal tunnels dug between shafts. During these works the navvies constructed the brickwork as lining to the tunnel, and small boys were used to construct the roof brickwork due to limitations on space. This would be somewhat against the present confined space regulations!

For many schemes, progress was often interrupted by technical and related financial problems. During the period 1805–9, a group of experienced Cornish miners, including

Richard Trevithick and Robert Vazie, attempted to dig a tunnel between Rotherhithe and Wapping, but failed due to difficult ground conditions. The Cornish miners were used to tunnelling in hard rock but in this case did not modify their methods for the soft clay and quicksand. Some years later Marc Brunel was to restart these works having just been working on a tunnel scheme under the River Neva in St Petersburg. First he turned his attention to designing and patenting a tunnelling shield which he successfully used at Rotherhithe in 1825. This involved the use of a 50ft-diameter iron ring to sink a large shaft. Thirty-six miners worked side by side in the shield in tiers of three. Further shafts were dug and two heavy plumb bobs dropped down them and steadied in tubs of water. A telescope was then lined up on their strings and the exact line of the tunnel calculated.

Irish labourers employed on the construction of the Kendal–Lancaster Canal in 1818, were attracted by lavish quantities of free liquor supplied by the local Conservative election candidate. When supporters of the candidate arrived, organised by Lord Lowther, the owner of numerous local collieries, the navvies showered them with mud and stones. It is likely that the navvies were more interested in having a good fight than any particular political motivation! Unrest was to also affect many railway projects. In the early 1820s, George and Robert Stephenson were engaged in jointly surveying the Stockton to Darlington railway. Resistance to various railway projects came from both landowners and workers. Farmers armed their own employees with pitchforks and guns. At St Helens in Lancashire one chainman was held by a mob of colliers who threatened that they would hurl him down a coal pit. In other cases surveyors were stoned and theodolites smashed. The reports of the Peterloo riots may have acted as a catalyst. Nearly twenty-five years later, Brunel's surveying assistants were also attacked by the landowner's groundsmen in the Forest of Dean area prior to the construction of the South Wales Railway link with the Great Western Railway.

During the 1830s there were many major railway contracts involving large-scale tunnelling operations. The Box Tunnel between Chippenham and Bath on Brunel's Great Western Railway scheme constructed in 1836 initially required six permanent shafts sunk into the hill. It was a tremendous task carried out under candle light, using the power of men and horses. Excavated material was brought up in buckets, hoisted by horses at the surface. For a considerable distance the tunnel passed through freestone rock. The results of trial shafts 'gave full assurance of the work being free from unexpected difficulties'. Unlike colliery works the danger of meeting explosive gases was remote. However, in 1837, Isambard Brunel was facing a disaster as the water flowed so freely from the rock fissures that the steam engine used to pump it out proved insufficient, and one division of the tunnel was filled. The water rose 56ft high in the shaft, making it necessary to suspend operations until the following summer when another steam engine was brought in to assist. The project was completed in 1841 at the cost of a hundred lives.

In 1837 at Clay Cross, Derbyshire, six shafts were sunk along the tunnel route. During the cutting operation for the railway a rich seam of iron and coal deposits was encountered, and George Stephenson & Co. became involved in sinking the colliery workings. They also built houses for the tunnel navvies and their families which was not always the case for employers at that time. For the Mersham railway line in 1839, another famous civil engineer, John Rennie, needed to arrange for borings and shafts. Like shaft sinking in collieries, tunnelling works were dangerous, but the 'navvies' were less disciplined than miners and appeared to take delight in taking unnecessary risks. However, the main causes of deaths and injuries were falls of rock and sudden rushes of water. In most cases there was no compensation from owners, although sometimes the contractors' agents who had been former navvies contributed a guinea towards funeral expenses. At the end of each contract, the navvies normally moved onto the next canal project, rather than going back

Group of navvies working on the Manchester Ship Canal in 1890. (Manchester Archives and Local
Studies)

to farming. However, in 1794 the start of work on the Peak Forest Canal was delayed until
after the harvest was in.

During 1841, near St Andrews, Bishop Auckland in County Durham, there were
frequent commotions due to gambling between colliers and navvies engaged on the tunnel
and railway construction. At Gyfeilliom (Hopkinstown) and Dinas in South Wales there
were sinkers at the new pits as well as hordes of hard-drinking navvies engaged on railway
construction works, always ready to resort to violence, if only for recreation. During the
Woodhead Tunnel works carried out between 1839 and 1845, a local GP, Dr Robertson,
was alarmed by the number of casualties. He contacted his friend Edwin Chadwick who
had been a commissioner dealing with employment of children in factories. This led to
a Government Select Committee being formed to investigate working conditions on
railways and public works. By 1848 there were 200,000 men working on the railways, and
often site conditions were very basic. A former navvy for twenty-seven years reported how
men were killed or injured when being hauled hundreds of feet up an air shaft. It was not
surprising, however, that nothing came of the investigation, given the pressure of railway
companies on Members of Parliament.

During the railway works near Lockerbie in 1846, running battles between English and
Irish navvies required the local authorities to call on the services of the militia. Not far
away a large-scale riot took place at Penrith, which started apparently after an Irish labourer
refused to take an order from an English foreman. For similar reasons a riot broke out at
Gorebridge near Edinburgh. Nevertheless, both the navvy and sinker at work normally
accepted the need for discipline, but outside work the bravado culture came into play.
There were navvies and shaft sinkers working on the Mickleton Tunnel, part of the Great
Western Railway. The Riot Act was read due to the long-running dispute between famous
engineer Brunel and the contractor Peto & Betts. This came to be known as 'The Battle

of Mickleton Tunnel', 1851, when Brunel's private army of 3,000 navvies fought the army of the disgruntled contractor who was backed by the forces of the local magistrates and armed police. This was said to be the last battle on British soil fought by private armies. The engineers were armed with pistols and cutlasses and the navvies' weapons amounted to pickaxes and shovels. Eventually George Stephenson was brought in to act as an arbitrator, perhaps the only person that held the respect of both engineers and workmen.

The construction of the Crystal Palace Exhibition in 1851 attracted labour mainly from Lancashire, Durham and Northumberland, and a few from Kent, Cornwall and Ireland. Shortly after, the Crimean War was to draw on the support of many sectors of the British construction industry. Apart from the navvies, many other tradesmen and specialist sub-contractors such as well sinkers contributed to the war effort. As with some coal owners, there were certain contractors who had some regard for their workers' welfare, for example, Thomas Brassey of Birkenhead, the famous railway contractor, whose navvies worked in Crimea. James Beatty, a railway engineer from Enniskillen who led the Railway Construction Corps, employed well sinkers as well as navvies. The Corps of Sappers & Miners installed the electric telegraph in the Crimea as well as their main function in the siege works about Sebastopol. They were retitled the 'Royal Engineers' in 1856. The navvies worked in four wars – against Napoleon (press-ganged), in the Crimea, Sudan and the First World War. In the 1860s former navvies were engaged in well sinking at Craven in Cleveland.

During the nineteenth century there were several attempts to tunnel under the Channel, including one by Napoleon. In 1877 an Anglo-French venture started with the sinking of a shaft to a depth of 330ft at Sangatte in France, but the English works had to be abandoned due to flooding. In 1880 near Dover, shaft No.1 was sunk with a horizontal gallery driven along the cliff at 10ft above high water mark. After Welsh miners had bored 800ft of tunnel, a second shaft (No.2) was sunk at Shakespeare Cliff in 1881, and then a tunnel was started towards a mid-Channel meeting with the French. Both tunnels were to be bored using a compressed air boring machine invented and built by Colonel Frederick Beaumont MP, a descendent of an early boring pioneer. However, a rotary boring machine patented by Captain Thomas English was used instead. Due to funding problems and political security fears, the project was terminated. The Channel Tunnel was eventually restarted in 1988, taking six years to complete.

Survival sometimes meant being able to react to the unexpected. In the 1880s on the Scottish Central Railway, a navvy had just lit a fuse at the shaft bottom and was ascending in the skip. The winding horse fell with the 'staggers' and the navvy immediately jumped out of the skip to snuff out the fuse just in time. As with mining, the death of shaft sinkers on canals and railways did not lead to much improvement in safer means of sinking, despite the Navvies Union formed in 1891 by John Ward who worked on the Manchester Ship Canal. In 1899, an accident occurred during the construction of the new Meon Valley Railway in Hampshire. One man died due to the collapse of the sides of the tunnel shaft, and his fellow worker was rescued after being buried in debris for over 50 hours. The sinkers on these works had to face similar dangers as colliery sinkers, but they were unlikely to have the protection of trade unions mainly due to the temporary nature of their work.

The census records of 1901, showed many navvies residing at Mrs Bennett's Lodging House in Sheffield. The navvies came from various British counties and a large number from counties of Ireland such as Leitrim, Mayo, Sligo, Tyrone, Roscommon and Fermanagh. Also recorded was a navvy and tunnel miner from Leicestershire. At Chepstow in Monmouthshire, Edward Smith from Carlisle was described as a 'tunnel miner navy' and vagrant. There is evidence that after working in coalfields, sinkers were able to use their specialised skills in other fields such as docks and bridgeworks.

2

Shaft Sinking Methods, Equipment and Contracts

Shaft sinking at its most basic is excavating a hole in the ground. In the case of finding water it amounted to digging until the water table was reached. However, sinking a pit could be fraught with dangers and difficulties, and these greatly increased costs as shafts went deeper. There have been various types of mining including the use of bell pits, drift and slope mines, shallow shafts as well as the later deep shafts, and open-cast mining. It was with the latter types of mining involving large investment that the need for exploratory boring became essential.

The importance of the shaft to colliery workers was acknowledged by their use of the word 'pit' as synonymous with colliery. It became a custom by owners in the Walker area of Tyneside, and generally in Durham, to name the pit shafts using the female first names (often in alphabetical order) of owners' relations; thus the first pit at Walker was named Ann Pit. In other coalfields, pits were named in a variety of ways. One of the pits at the St Helens Colliery in Lancashire was sunk in honour of the Princess of Wales. It was named Alexandria Pit and the princess visited the colliery during the sinking.

The treatise *De Re Metallica*, produced in 1556 by Agricola, essentially described the most advanced techniques in drift mining, principally for metals, as practised in mid-sixteenth-century Germany. His recommendations were addressed to mine owners or investors. He noted that when a vein below the ground was to be exploited, a shaft was used and a wooden shed with a windlass placed above it. Methods for lining tunnels and shafts with timber were described, and listed were basic hand tools, lifting machines, some man-powered and some powered by up to four horses or by waterwheels. Horizontal drive shafts along tunnels allowed lifting in shafts not directly connected to the surface. Several designs of piston force pumps were included and powered similar to the lifting plant. The designs of shafts for ventilation with wind scoops, fans and bellows were also described.

About 150 years later and before steam power impacted on mining, a writer with the initials J.C. from the north-east of England produced an important manual of best practice for pit sinking and winning of coal titled the 'Compleat Collier'. This manual recommended the use of exploratory boring in advance of the sinking operation. It also highlighted the dangers of explosive gases that might be encountered underground. The author emphasised the need for the sinking work to be controlled by experienced sinkers and the importance of good communication between owner, viewer and master sinker.

Writing at the same time, Daniel Defoe painted a picture of the process of obtaining coal from the mine to the consumer:

They are dug in the Pit a vast depth in the Ground, sometimes fifty, sixty, to a hundred fathoms, and being loaded, for so the Miners call it, into a great basket or Tub, are drawn up by a wheel and horse, or horses ... to the top of the shaft or pit-mouth and there thrown out upon the great heap, to lye ready against the Ships come into the Port to demand them.

The main types of sinking are described as follows:

Bell Pits

Bell pits were shallow bottlenecked pits found excavated in many mining areas. They consisted of vertical shafts down to the excavation below and were usually circular but occasionally oval, and rarely exceeded 20ft in depth. Having reached the limits of ventilation or the miners' ability to support the roof, the shaft was abandoned and a new shaft sunk the least possible distance away. The spoil material was used to backfill the abandoned pits. The increasing difficulty of draining the pits was another reason to open up a new excavation. In the early years, it was easier to dig new bell shafts than to construct underground roadways. It was generally cheaper to sink shallow shafts than to transport coal more than a few yards underground. Ladders were fixed to the pit sides and coal was brought up to the surface in baskets. Later the baskets were brought up on ropes by means of a windlass powered manually or by horses. Often this type of mining was carried out part-time with farming, for example in lead-mining areas. Outputs of about 15 cwt to a ton per man-shift could be expected. In Derbyshire and in Yorkshire, bell pits were used to work ironstone, spaced at about 20 to 40 yards apart.

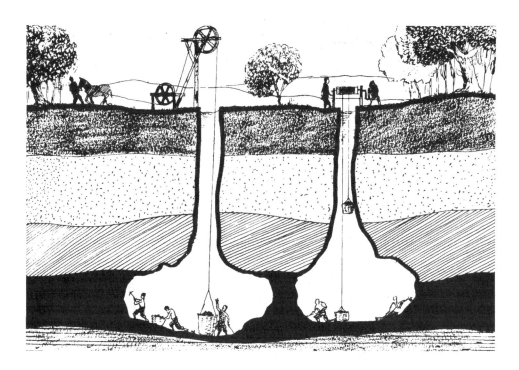

Undated illustration of bell pits, one pit with whim-gin and other pit with windlass. (Keele University Library)

Deneholes, a primitive form of excavation, were a type of bell pit dug to obtain chalk or marl used for fertilising fields. The name may have been derived from 'Dane hole' as used by Danish invaders. The chalk was excavated using a short-headed iron pick, working forward in a series of steps or benches. These abandoned benches have led to stories that these excavations were used for druids' altars. Initially the shaft was about 3ft in diameter and about 20ft deep, and footholds were cut in opposite sides of the shaft so miners did not require a ladder. The chalk was hauled to the surface in a basket using a surface-mounted windlass. Hazel basket-work was used to prevent the sides of the shaft falling in. These pits were often backfilled by throwing tree-stumps and bushes down the shaft and then filled to the surface. The deneholes were dug near field boundaries to minimise problems to farming if subsidence was to take place around the pits. Later these chalkwells consisted of 4–6ft-diameter shafts sunk down to roughly cut caves radiating from the base of the shaft. Shafts have been found sunk as deep as 300ft in the chalk beds of northern France and southern England, using tools such as wedges and picks made from deer antlers and shovels from oxen shoulder blades.

Drift Mining (and Slope Mining)

Drift mining was the most basic form of mining after open-cast methods. The scarcity of wood resources encouraged a change from bell pits to drift mining to increase output, although in some areas drift mining developed first, and in some cases bell pits were never used. Drift mining or 'drifting' consisted of a rough form of tunnel or near-horizontal adit into the shales from the side of a hill. Slope mines differed from drift mines with the former having an inclined entrance from the surface to the coal seam. If possible, drifts were driven at a slight incline so that removal of water could be assisted by gravity. In the early mines, the adits or drains which carried off the water served also to admit air. The noxious gases, whose quantity in the higher seams was not great, were carried by the air and passed up the shaft or out of the adit. When the mines were increased in depth and extent, the risk of explosion greatly increased. In Scotland the driving into the sides of a hill was known as a 'creeping heugh' and later by means of vertical pits called 'windlas heughs' drained by adits. Where drains crossed over a neighbours' land, compensation payments would normally apply.

Shallow drift workings known as 'day-hole' pits were used near Heckmondwike in Yorkshire. The name indicated the short-term nature of the excavation. Some were large enough for access by men and horses. Often the old entrances to drift mines were rented out in a few fields for coals. The tools used in the 'day-hole', bell pit or drift were nothing more than a few barrows, some iron nails, and a hand-winch or gin to draw the materials to the surface. These simple methods of working were in some areas continued into the 1850s. At Dowlais in South Wales at this time, the winding gear was worked not by water or horses but by girls!

In the mining of minerals such as lead, the most basic operation consisted of following the vein along the surface and digging it out, much in the way of open-cast mining. In some mining there was a combination of drift mining and shafts. In Cumberland, men, horses and materials were accessed through a 'bearmouth' – a local term for a tunnel within the seam, opening to the surface at the outcrop, and dipping into the mine and down the incline of the seam. Rather than dragging containers through the galleries, short shafts were sunk down to the seams and the containers were wound up using hand-powered windlasses or horse-powered gins. The shafts were carefully located slightly up the incline from the level, so that if they were sunk further to a deeper seam, these deeper workings were not at risk of catastrophic flooding in the event of any back-up of water in the level.

Shaft Sinking (Deep)

The first stage of sinking required 'cutting the sod' which was frequently celebrated with an official ceremony, sometimes with religious blessings. 'Groundbreaking' came to be known as having special significance, and often the actual spade or shovel used in the ceremony was kept for subsequent display. Topsoil was then removed over the working area for reuse.

As described by the Compleat Collier in 1708, the northern method of sinking began with the cutting of a four-sided pit cut into the surface soil. As sinking proceeded towards the stone it was shaped to an octagon, and the shaft through the stone itself was circular in plan. The sides above the stone were timbered with fir baulks and lined with deal boards to prevent falls of earth. When sand was encountered, it was usual to hold it back by ramming clay between it and this wooden tubbing. When the miners came to wet strata they sometimes packed undressed sheepskins between the boards and the stone, and then the sides were lined with bricks, behind which spiral channels, known as garlands, were constructed to carry to the pit bottom water which would otherwise have forced its way into the shaft. On completion of the shaft to the seam, a sump was further excavated to

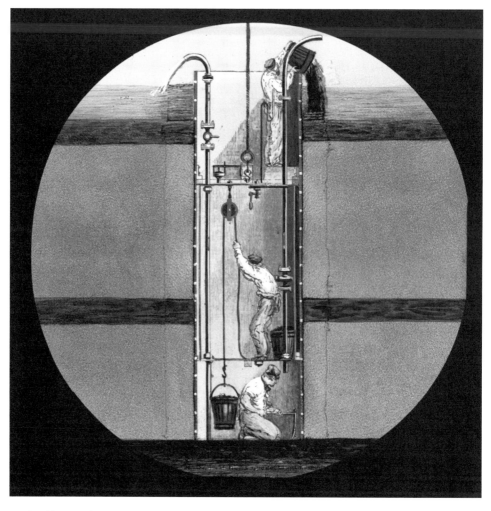

Undated lantern slide showing shaft sinkers in shaft excavating with pick, winching up excavated material and pumping out water. (Beamish Museum)

provide 'standage' for water. Often there was a need to sink another shaft alongside to drain away water. The shaft was 6ft 9in square, and with deduction for timbering, leaving a diameter of about 6ft for the finished pit. The sinker only used hand tools such as a mattock, a gravelock (a stout iron bar), a sledge hammer, and several short wedges. Smiths were on hand to keep the 'hacks' sharpened. Later, when solid strata were met, gunpowder or other explosives were used to increase progress.

The location of the main shaft was important and was based on ground investigation, and later including the use of boring. Staple shafts were sunk to provide access between coal seams. The proposed method of working the coal could influence the shaft location depending on the size of the colliery 'estate'. Sinking in rural areas gave better returns, since there were fewer buildings to remove and it minimised the amount of coal that had to be left for stability reasons. At earlier times, the contour of the ground surface was taken into account when balancing between setting up unnecessarily at a high point, whereas a low level could have flooding problems. The slope of the strata, known as the 'dip', and its compass bearing, the 'strike', were used to estimate location of seams.

The danger of stythe or bad foul air combined with the use of candles could cause explosions, and as well as killing men might also lead to destruction of the colliery by letting in water feeders. Roadways were driven through the seams where there was least resistance, with 'drifting' through stone strata only when necessary. The early single shafts usually had an oval cross-section as this gave more space for ventilation, but as shafts went deeper a circular cross-section was used since this was better at resisting the increasing earth pressures. While in Cumberland and sometimes in South Wales there was a preference for oval-shaped shafts, for coal mines in the Newcastle district the shafts were invariably circular, 7–8ft in diameter. Shafts with larger diameters were sunk for accommodating pipework and should an extension of the shaft be necessary. As shafts became deeper the shaft diameter increased, as at South Hetton colliery where the shafts were 15ft diameter divided into three parts from the centre.

Access for sinkers and miners up and down the shaft was initially gained by means of a rope, knotted at regular intervals. Later, corves or wicker baskets were then tied to the rope. This arrangement was superseded by chains and iron containers. Sinkers descended the shaft sitting astride kibbles (large iron buckets also known as 'kettles' or hoppets), each man with one leg in and out of it. Working platforms called 'cradles' or 'scaffolds' were erected in the shaft, using the kibble to supply and remove materials and tools. In the North East little protection was provided above or below the cradle.

The change in use from metal tubs to cages in the shaft reduced the damage to the shaft lining caused by the tubs crashing against the sides of the shaft. It was important to achieve a near vertical shaft to avoid operational damage to the shaft lining. As shafts got deeper the accuracy of the positioning and alignment of the shaft became more critical. At Barnsley, Silverton Colliery, sunk in 1860, a plumb was used for the entire depth to ensure verticality. Sinkers were thus able to measure from this line to achieve a constant diameter. The walls of the shaft were constructed as the shaft was sunk; these walls made of brick or stone were thicker at the top than at the bottom section.

Absenteeism amongst hewers could be as high as 10 per cent, due to the exhausting nature of the work. In contrast, sinkers often worked with barely a break until the shaft reached the coal seam, and then did little work until the start of the next sinking, which could be delayed due to various circumstances. At earlier times, sinkers' clothing did not provide much protection, especially to the head, and sinkers were very vulnerable to the fall of heavy objects from above. Sinkers wore leather hats called 'backskins' with a large flap (backflap) at the back to prevent water drips going down the neck, similar to those worn by fishermen. They also wore leather coats to protect their shoulders and backs from dripping water and small debris falling down the shaft. In the wet collieries found in many parts of Scotland, putters sometimes wore 'backskins' to deflect the water and as protection to abrasion. When not wearing the helmet, the sinkers wore a leather skullcap giving little protection. Tallow candles used by early sinkers and miners, sometimes fastened to their helmets, were cheaper than wax but smellier.

Communication between the sinkers and engineman was of vital importance. In the report of the Horden Colliery Disaster, the practice of ascending and descending by the kibble involved giving signals by hitting the side of kibble with a hammer:

one rap meant – 'hold' or 'stop'
two raps – 'to lower'
three raps – 'bend up'
four raps – 'bend away'

It should be noted, navvies, and probably sinkers, made very little noise when working, aware that all their energy was needed for the task in hand. It was said that navvies employed on canals and railways worked in silence in order to conserve their energy. However, noise from equipment could be deafening. The banksman worked at a heppstead 20ft above ground level, and had a duty to transmit any signals from the shaft to the engineman. Of course the means of communication was by voice, and the master sinker would shout his instructions to his sinkers down the shaft. My great-grandfather James Mason, a master sinker, apparently had such a strong voice that it could be heard as far away as the colliery houses!

A great improvement in the method of sinking shafts through water strata was invented by Messrs Kind and Chaudron, and the shaft at Whitburn was sunk in 1881 using this technique. The system consisted of boring out the shaft from the surface until the watery strata had been pierced, and a suitable foundation obtained on which to place the cast-iron tubbing. The tubbing, consisting of rings the full size of the shaft, was then lowered into position, and then by means of an ingenious stuffing-box arrangement at the bottom, packed with moss, a watertight joint was formed as soon as it rested on the bed prepared for it. The water in the shaft was then removed, and then sinking in dry measures was then carried on in the usual manner. Towards the latter part of the nineteenth century, there was a need for larger shaft diameters and further modifications in sinking techniques. At Astley Green, Lancashire, in 1908 a dropshaft or sinking drum process was used. This consisted of forcing cylinders of iron, with a cutting shoe at the bottom, through the ground with hydraulic jacks and excavating within it. The next section of tubbing was then attached and the process repeated.

By the twentieth century sinkers were able to take advantage of progress in mining technology, including freezing and cementation processes with considerable impact on inflows during sinking. In the long run these techniques made sinking both quicker and safer. Three shafts were considered desirable, one for pumping and the others, downcast and upcast, for ventilation. At the beginning of the twentieth century mine inspectors were starting to exert more influence both on the capability of machinery as well as safety.

With today's technology, the modern design of shafts is based on geotechnical and structural considerations. This would be carried out by professionally qualified mining or civil engineers. The first methods of tunnelling were simply adaptations of those used in shaft sinking. Since tunnelling as a rule presented greater ground-water problems than shaft sinking, the application of grouting to rock tunnels seems to have developed rather more in civil engineering practice than in mining. For example in shaft sinking below dams, grouting was used to strengthen the rock and to prevent inflows of water during driving, and also to prevent outward leakage under the great hydrostatic pressure to which the tunnels are subjected to when in use.

Group of pit sinkers at South Moor Colliery, Charley Pit, Durham, in 1892. (Stanley Library)

A further group of pit sinkers at South Moor Colliery, Charley Pit, Durham, in 1893. (Stanley Library)

Summary of Sinking Methods

The methods adopted for boring and sinking shafts were determined by the nature of the ground including the level of water feeders. These methods developed in sophistication through the nineteenth and into the twentieth century. The quality of ground information already available was crucial to the method adopted.

In 1909 Richard Redmayne, in his book *Modern Practice in Mining*, set out the options in shaft sinking in varying grounds:

1) Ordinary & fairly dry strata:
Hand excavation at bottom of shaft with blasting and drilling through stone; Temporary wooden cribs with backing deals replaced at lower levels by brick or stone walling; Drainage by means of water kibbles.
2) Water bearing strata:
As in 1) with water pumped and shaft lined by coffering or cast-iron tubbing.
3) Thick surface deposits of sand, gravel or clay:
Larger diameter shaft and driving wooden piles, tier within tier, or interlocking metal piles or Walker's Method of driving steel piles carrying tubbing.
4) Thick surface deposits of sand & gravel:
Hoase's System consisting driving tubes side by side forming a shaft lining.
5) Surface loose water-bearing strata:
Sinking-Drum Method by which a 'drop shaft' of cast-iron rings is lowered from surface as excavation proceeds.
Sark-Borer Process by which shaft is bored out and debris in sacks winched out.
6) Surface loose water-bearing strata limited to a depth of 120ft:
Pneumatic System of M Triger – caisson lowered as excavation proceeds with water excludes by compressed air.
7) Surface sands, gravels & clays:
Honigmann System – boring out shaft and applying muddy water under high pressure to maintain the shaft sides and raise debris.
8) Thick deposits of sand & gravel or when penetrating rocks which are heavily fissured and contain water:
Cementation Process – ground consolidated first by injection of liquid cement, and then sunk by ordinary methods.
9) Firm strata containing large quantities of water which cannot be coped with by pumping:
Kind-Chauldron Process – boring out of shaft, lowering down complete rings of metal tubbing, and afterwards pumping out water.
Hydraulic-Ram Boring Process – boring out with force also applied by hydraulic shock to a series of cutters at the bottom of shaft.
Pattberg System – similar to Kind-Chauldron with debris forced by compressed air to surface.
Dropshaft or Sinking Drum process – a process similar to Kind-Chauldron involving forcing cylinder into ground and excavating within it.
10) Loose water strata either at the surface or when occurring at some lower depth:
Freezing Process – water-bearing strata is frozen in the vicinity of the shaft, then shaft sunk and lined as normal.

Contracts and Legal Conflicts in Sinking Shafts

Royalty in earliest times granted rights for mining of coal, and the Church for many centuries had a direct interest in determining the restrictions on the coal trade. Gradually these rights were transferred to other bodies. Disputes often arose from conflicting interests between colliery owners and proprietors who found it necessary to seek both technical and legal advice from engineers and sinking contractors.

Apart from rival claims to the rights of ownership and working of minerals, there were associated issues such as the value of minerals and concessions, and permission and cost of using wayleaves. Also in the area of excavation and workings there was a cost relating to the detriment to the surface value. Considerable tracts of farming land might have to be sacrificed for the sinking of shafts. Also the effect of underground operations could cause ground subsidence leading to compensation claims. A contract placed obligations on all parties, but sinking contractors were aware that the terms and conditions of the early contracts were strongly weighted in favour of the colliery owner. These contracts included little reference to working conditions or health and safety. In some southern counties, individual miners struck 'bargains' to determine wages based on the market price

Sinker at pit head gear of Wallsend G Pit, Rising Sun Colliery, Northumberland, in 1892. (Beamish Museum)

Shaftsmen preparing to descend the shaft at Wallsend H Pit in 1894. (Beamish Museum)

of the mineral at that time. The following records give examples of some contracts and the difficulties which arose during mining and sinking works which sometimes required specialised technical and legal resolution.

From the thirteenth century the lead miners at Alston Moor in Cumberland were given legal privileges, and by the reign of Henry V a 'court of mines' came into being, with the king's officers no longer able to serve writs. Derbyshire in a similar manner was divided up into many small 'liberties' for the purposes of lead mining. Despite these legal arrangements, many disputes arose over boundaries and the rights to water and timber. The drainage of mines in close vicinity to other workings and habitations led to the introduction of the 'Laws and Customs of the miners in the Forest of Dean'. In ancient times the bounds of a mine had been fixed as the distance to which a miner could throw his spoil from his pit. Using this law, other miners started to sink their pits so near as to rob the benefits of an existing adit. It was therefore necessary for the Mine Law Court to enact in 1678 that no miner might sink a pit within 100 yards of an adit belonging to another man without his consent. This distance was extended in 1692 to 300 yards, and subsequently to 1,000 yards. Similar restrictions were also applied to the use of boring.

In mining operations it was precarious to depend on a neighbour's pump. In the Mendip region in 1617, Richard Burke introduced a 'mill wheel ... to drawe the water' from his small diggings in Midsomer Norton; his competitors, James and Clement Huishe and others, 'so worked underneathe the earth that they caused the water to runne into [Burks's mine] ... whereby ... the mill wheele might drawe the water of both.' Instead, they 'utterly drowned' the works. Thomas and John Briggs, undertakers of the great Bedworth colliery in Warwickshire, were charged with flooding the pits of their rivals

at Grill. It was considered that the latter only had themselves to blame, because they 'depended for drainage solely upon the two engines' erected at Bedworth. They were without resources of their own, and therefore in trouble when the Bedworth partners, finding fresh coal nearer the surface, allowed the water to stand fourteen 'ells' higher in the pits'. In Warwickshire in 1623 the lessees of Bedworth coal mines were prohibited from boring any holes which might endanger adjoining mines by flooding them with water let down from the surface.

At Killingworth, Northumberland, in 1762, the viewer's requirements for sinking of the coal and engine shafts were recorded in his notebook as follows:

> Order'd that a drift be drove towards the tube staple [sic] immediately & that the same be sett out of the shaft at right angles from the water levell drift and an the same levell & then to be turnd in a line toward the tube staple ...The sink.s [sinkers] propose to sink ye engine pitt to the 30 inch coal at £5.17 p[er] fathom cutting out crib holes taking down ye pitt & rough cladding included.The Owners to find a horse.This to be considered & ye agents to be aquainted therewith. It seems to deserve abt. £5 or Guineas.

The following day, the sinking using the Newcomen engine was agreed with the sinkers, on the basis of the shaft being 7ft 9in diameter at £5 5s per fathom with the owners supplying the horses for drawing the stones. It is clear from other recordings that many other trades were involved such as smiths or engine-wrights, ropers and carpenters supplying cylinders, plates, pistons, brass rims, bellows, flanges, bolts, beam catchpins, bottom and clack pieces, ropes and chains, and timber gates.

Into the nineteenth century, conflicts between adjacent colliery workings often continued to be very protracted. In north Derbyshire, by a letter dated 1821, John

Side view of shaft with sinkers fixing pre-assembled wall units at an unknown colliery in 1896. (National Coal Mining Museum)

Carleton and the Smiths were accused of illegally taking coal from 'the Liberty of Staveley Upperground Colliery, from the Old Level' which bordered the Arkwright estates. This was denied by the Smiths, claiming that they had been given permission to mine by the Duke of Devonshire's former land agent. A letter dated 1832 confirmed that the lease of 1819 between the Duke and the Smith company for mining the Staveley Colliery and the Hady ironstone mines had been cancelled. Smith & Co. were eventually ordered to pay compensation of £752 7s.

In Merthyr Tydfil the ironmasters were often the magistrate for offences brought against them by their workmen. In 1858 a sinker called Ben Hodgkiss went to Madeley County Court in Shropshire to claim for non-payment of wages amounting to £1 4s. The defendant was a German engineer, known as Joseph Reider (real name Donster Swivel Redivivus), who had been engaged by Mr Moseley of Buildwas Park. Dr Reider proposed to sink for coal in a location where Government surveys and scientific experience had indicated that coal would not be found. The plaintiff stated he was a sinker working for 4s a day and two quarts of ale. On Christmas Eve he had been given notice to quit on the following Saturday. His case was that he had been unable to work on Monday, it being a wake day, and that he worked longer than normal hours the next day to finish four rings of brick [lining] to allow men to work the following morning. He had claimed a week's wages but had only been paid for a day. The defendant complained that Hodgkiss had 'behaved badly in the shaft', cursing him and Mr Moseley, and was more often drunk than sober. The judgment was given in favour of the defendant on the basis that Monday was not a general holiday, also taking account of the sinker's abusive manner.

Above left: Undated photo of pit head winding machine. (National Coal Mining Museum)

Above right: James Mason, master sinker and the author's great-grandfather. (Author's collection)

Below: Pit headstock at an unknown colliery. (National Coal Mining Museum)

Sinker pre-fixing timber circular cribs to shaft lining at unknown colliery, undated. (National Coal Mining Museum)

Sometimes there were disputes between mine owners when their respective working areas overlapped. In 1864 at New Bowson Colliery, located in the Forest of Dean, Gloucestershire, the Great Western Deep Coal Co. (GWDC) commenced sinking two shafts close to Winning Pits belonging to Messrs Goold & Heyworth. Two years later, the shaft brick lining was in the process of being replaced with substantial stone, however, the lining was still being forced out. A piece of stone fell causing a fatality. To avoid further dangers, as directed by the Mines Inspector, sinkers started fixing deal boards over the offending areas. In order to reach their allotted coal, GWDC had to sink their shafts through ground already granted to Goold & Heyworth. The sinkers were met by a sudden inrush of water, and the only remedy available to remove the water was by hauling it up the shaft in wooden containers requiring two large engines. When management realised that de-watering by this method was not viable, they purchased a Cornish beam pumping engine from a recently closed mine at Poldice in Cornwall. A new company was formed and a staple pit was sunk to receive the second-hand pumps. The double-acting pumps closely followed north country practice, where the engine would draw up water from the bottom to a mid-point in the main shaft from where it would flow to the staple pit. The foundry which rebuilt the pumps, as in the tradition of erecting Cornish engines, would

Shaft sinking at Cwm Colliery, Glamorgan, in 1909. (West Glamorgan Archives)

Group of sinkers at Blackhall Colliery, Durham, sunk in 1909. (Beamish Museum)

have supervised the building of the engine house. It is believed that a fight took place below ground. The ensuing court case found that the new workings posed no threat to Goold's since GWDC had an engine capable of winding 5,400 gallons per hour, whilst a new engine was being erected with a capacity of 30,000 gallons.

A similar disagreement took place at Neston in Wirral Cheshire, when close interface between coal lessors' working areas led to subterfuge by their workers. Ness Colliery, which opened in 1759, was owned by the Stanley family, and coal was exported to Ireland and North Wales. They soon recognised the need to extend their underground workings north-westwards in order to follow the coal seams as they dipped below the foreshore and under the adjacent manor of Little Neston. Even though Little Neston coalfield was owned by the Cottingham and Talbot families, they allowed Stanley to sink new shafts and extract coal on the foreshore of Little Neston providing Stanley paid them a share of the profits arising from this agreement. Relations broke down in 1819 when the rival companies were working in the same seam only 200m apart. Two years later Sir Thomas Stanley's men descended a shaft at Ness Colliery armed with explosives, pick axes and other equipment. They then moved into the workings of Little Neston and with the use of explosives blocked the underground road which was vital to Cottingham's operations, rendering this area dangerous to enter. In 1821 Thomas Cottingham brought a lawsuit against Sir Thomas Stanley for 'a trespass, and injury done to his property'. The Stanleys were ordered to pay the Cottinghams the sum of £2,000 in damages. However, this did not end the bitterness, when in 1829 Stanley was willing to sacrifice a seam in order to flood Cottingham's works.

A story from the West Midlands suggests that sinkers with an early knowledge of the outcome of the sinking were sometimes bribed by colliery officials. At the turn of the twentieth century shafts were being sunk at New Hawne Colliery in Worcestershire, and an old beam engine was erected. However, for no obvious reason the shafts were not worked. Rumours started to circulate that a local mining engineer was allowed down the shaft after 'tipping' the sinkers, and after seeing the quality of the coal newly exposed, bought the adjacent mine. Another explanation put forward for the standstill was that the engineer had considered the quality of the coal to be inferior to that being produced by other local collieries and therefore would not have been competitive.

In 1883 the men at Wardley in Durham refused to work in the pit and remained out for two days while a fire was extinguished. One month later it was discovered that the fire was still burning and had moved to the other side of the ventilation furnace from which it had started. The men were angry since they believed they had been kept in ignorance, putting their lives in danger. The fire was 80 yards from the 'goaf' (the name for exhausted workings) where there was an accumulation of gas and close to the 'dumb drift' through which came all the bad air from the workings by the ventilation machinery. The men contacted the sub-inspector, Mr Atkinson, who told the men not to return to work until they themselves were satisfied that the fire was out or to send for him to decide. Although it was not unusual for men to be prosecuted for endangering lives, this showed an early example of miners getting redress when they saw the management at fault.

Contracts

It is interesting to compare the contract or legal restraints which have applied to the work of miners and sinkers through the ages. Firstly, the 'bond' was a heavy legal straitjacket which reduced the life of a miner to that of a serf. The obligations of his master were very few.

At earlier times, the contract would have been a verbal agreement sealed with the shake of hands. The contractual restrictions on the sinkers' work at the beginning of the

Above: Undated photograph of a group of shaft sinkers at Waterhouses Colliery, Durham. (Beamish Museum)

Left: Shaft skip pocket under construction at Crown Shaft, Westoe Colliery. in 1909. (Beamish Museum)

eighteenth century were set out in the 'Compleat Collier'. Two centuries before, the cost of purchasing mining equipment, of digging a pit to the seam, and perhaps a short trench to drain off the water from the working place, rarely exceeded £15, while less than £5 often sufficed. Sometimes beer money was added. In these contracts there was a lack of mention of health and safety or any factor that would increase the cost of winning coal. The works in shaft sinking involved a number of trades other than the work by the sinkers, such as works by masons, joiners, smiths, specialists for chimney builders and ironwork, and horse hirers. Works such as cutting foundations and drains may have been carried out by sinkers, or labourers employed by sinkers.

The Conditions of a Lease to the Marquis of Londonderry of the Coal Mines at Bishopwearmouth in 1847 included the following counter clauses:

> Lessor or his viewer to have the liberty at all times to descend by the machinery of any of the pits belonging to the Lessee, to survey and inspect the workings in the said coal agreed to be taken & ascend the same pits again.

Sinkers completing shaft sump at the bottom of the main winding shaft at Brayton Domain Collieries No. 5 Pit, Aspatria, Cumberland, in 1910. (Beamish Museum)

Sinkers with drills at shaft top in 1910 at an unknown colliery. (National Coal Mining Museum)

Lessee to have power of shaft way leave, outstroke, instroke, air courses & all other underground privileges that he may require for working adjoining coal.

The Cost Book for Wheal Agar, Cornwall, 1855–9 gave details of the works and agreed payments for various sinking contractors. The cost of sinking varied greatly determined by ground conditions as sinking proceeded. At Towneley in Durham (1859), the sinking of the Stargate shaft included drawing and 'banking out the Stones' as well as cleaning out the furnace. For the smaller concerns, the sinking contracts were drawn up to a simple format. For example, The Pit Sinking Agreement 1871 between the Derbyshire colliery owners Clay Cross Company and the sinking contractors Isaac Ward and John Smith consisted of an agreement and specification. At Stargate Winning in Durham, labour costs for sinking works in 1872 included works for shaft, winding engine and house, boiler seats and house, heapstead and screens and 'guides re Bruntons cages' (see Appendix 2).

The contract specification for sinking two shafts at Crawcrook New Winning on the 24 March 1890 ensured that certain technical requirements were met; for example, in the measures to be taken in blasting and excavation in order that a consistent and accurate shaft diameter was achieved. In many sections, it required that the contractor met the satisfaction of the company (or its engineer or agent). This was a 'catch-all' statement which could be used by the employer for his benefit. (This has been widely used in civil engineering specifications into modern times.) The balance of risk is weighted heavily in the company's favour, with clauses such as:

not responsible for correctness of geological data (no opportunity for the contractor to have any prior knowledge unless he had previous experience of the area).

not liable for any accidents.

effect of flooding to be determined by the Engineer.

termination of contract decided by Company.

Plan of shaft with sinkers fixing the timber lining at an unknown colliery in 1910. (National Coal Mining Museum)

Above: Sinkers in shaft constructing brickwork lining at an unknown colliery. (National Coal Mining Museum)

Left: Group of shaft sinkers at Westholme Colliery, Durham, in 1911. (Beamish Museum)

Group of sinkers working for the Ashington Coal Company at Ellington Colliery, Northumberland, in 1912. (Beamish Museum)

Undated photograph of sinkers in a kibble at an unknown colliery. (National Coal Mining Museum)

The specification clearly divided the responsibilities of each party to the contract which in the past may have been assumed according to custom:

supply of labour and equipment.
maintenance of tools and workings.
provision of housing and coal.
rates of pay including 'reductions' and 'advances'.

The company required that the contractor was responsible for the hire and full-time supervision of its sinkers, banksmen and labourers. This implied that there would be a master sinker in place but did not specifically use the words 'master sinker'. The contract allowed the company to get rid of certain workmen of dubious character employed by the

Above: Undated photograph of a group of sinkers at an unknown colliery. (National Coal Mining Museum)

Left: Group of sinkers at Barnburgh Main, South Yorkshire, in 1913. (National Coal Mining Museum)

General view of sinking towers at Vane Tempest Colliery, Seaham, Durham, in 1936. (Beamish Museum)

contractor. This may have been included to enable the company to remove men involved in union or strike activities. (See Appendix 2 for further details on all of the above contracts.)

Wages

The wages paid towards the end of the eighteenth century gave an indication how each workman was valued by the manager. A viewer was paid 15s to 20s a week, but as a skilled surveyor who could take three and four pits under his charge, it was not an excessive wage. The overman, who organised his men, and who was actually in charge of the daily operations, drew only 8s a week. The corver repaired the large wicker baskets in which the coal is drawn to the surface. The corves were damaged as they travelled up and down the shaft, and it was important that their capacity was not reduced by wear and tear. Banksmen were paid about 15d a day, and the barrowman who hauled the corves to the shaft was paid 22d a day.

In 1845 sinking at Selston, Derbyshire, was contracted at £12 1s per yard. A master sinker was paid £2 8s, compared to blacksmith at £1 16s and corver at 7s. The cost of sinking the pit was £18 16s 4d, thus the master sinker's cost was over 15 per cent of the sinking bill. By custom, many contracts included a daily allowance of beer, if not actually stated in the written contract.

By 1856 at Butterley in Derbyshire, surface workers were making 13–14s per week, hewers between 15s and £1 and enginemen from 15–18s. With deeper and larger mines, the sub-contract or butty system was abandoned after 1850 in most parts of the country, but still lingered on in the small and shallow pits of the Forest of Dean, Shropshire and the Black Country. Compared with other counties, the miners in the North East pits were better paid, and sinkers were always paid more than miners.

3

Transportation and Accommodation

The coal trade was dominated by transport of the mined material, but colliery officials paid little regard for the transport of miners. There was a steady improvement in the movement of coal starting in the late eighteenth century, from the use of pack horses to horse-drawn wagons, then from canals to railways, but the means of travel and accommodation of the miners and sinkers changed very little.

Colliery owners recognised that canals provided a much cheaper means of transporting coal, but weather could be a serious problem. During the winter period the canals would freeze over, and during hot spells a drop in the water levels prevented the passage of canal boats. Inland, the construction of canals in coal-mining areas started to make a big impact on the transport of coal. This was especially the case in the north-west of England with coal wharfs provided on the Shropshire canals. In 1798 the narrow canal between Merthyr and Cardiff was opened, and coal barges replaced the weary pack ponies. Although there were a number of proposals, no canals were constructed in Northumberland and Durham.

The system of railways connecting the collieries to the shipping wharves on the Tyne and Wear was in advance of any other area in Britain. A new type of staith was introduced at Benwell in 1807 by a Tyneside engineer, William Chapman. Keelmen carried out the arduous task of moving coal from the staithes to the collier boats until the first steam-powered paddle tugs were introduced in 1822. Adam Smith estimated that about a third of British shipping was employed transporting coal from the north-east of England to London. The surge in the coal trade stimulated shipbuilding, especially the coal-carrying colliers. In 1852 the first steam collier entered the Thames having sailed from Newcastle in 48 hours, consuming 8 tons of coal with a cargo of 600 tons of coal unloaded in one day. When adverse winds delayed the Newcastle coal convoy, important sections of industry in London were brought to a standstill. There were records in 1799 of sailing ships from Jamaica with coal loaded from Newcastle. All of Captain Cook's ships, *Endeavour*, *Resolution*, *Adventure* and *Discovery*, had been built and used for the coal trade.

The development of steam locomotion for the coal industry was being resisted by the keelmen, when riots were recorded at the new Sunderland Drops in 1815. The earliest steam engines used in the North East were stationary winding engines on hillside inclines, for example at Washington Moor. Locomotives, the 'steam engines on wheels', were first operated at Hetton Colliery near Houghton-le-Spring. Engines were first used for the purposes of raising coal in 1753, and such a device called a 'Menzie' was operated near Washington. The ancient method of bringing up coal to the surface by means of wicker baskets called corves was changed in 1833 by Mr Hall of Greenside who introduced the use of iron tubs at South Hetton Colliery. He afterwards brought in the use of cages to carry the coal-carriages up to the surface.

In 1841 the Children's Employment Commission noted that the young miners often access the shaft 'by riding upon the knees of anybody in the loops, the men, or big putters' or by corf, six at a time. A commissioner gave the following account of the risks of travelling in the shaft:

> Men went down together. The loops have hooks and are hooked into the link. We have one loop to two of us, one puts in his left leg and the other his right leg and with our hands we cling to the chain; frequently other two persons hang on the rope or chain holding by the bands and the feet, like sailors on the rope of a ship. Accidents sometimes happen. About four years ago at the Elvet Pit [Durham], when two men were going down a hook of another rope caught him by the hough, and ripped off the skin like a stocking. He got down to the foot of the shaft and then drew him up again, in the same condition as he went down and then they sent for the surgeon. The surgeon came to his house and sewed up the leg but it all became dead flesh and the man died five or six months after.

In 1843 at Seaham Colliery, pit sinkers travelled to work riding wagons on the Braddyll railway, named after the coal owner, but generally miners and sinkers travelled to the pit by foot, and then often for long distances underground. The wagons would have been needed to carry heavy equipment. Sinkers 'rode' the shaft in early times by the primitive method of clinging onto a rope or chain. The first known recorded use of pit ponies in Britain was in a Durham coalfield in 1750, but were generally introduced to mines in the early nineteenth century to minimise the amount of human labour required to move cut coal underground; that is, the work carried out by putters. The ponies were known as 'galloways' in Durham since many of the ponies came from Galloway in south-west Scotland. Initially ponies were taken down the shaft by means of nets, and later in a special harness within the cage. Once down the mine, pit ponies were rarely taken up to the surface other than at the end of their working life, or else as a carcase after injury. Dead men, boys and horses were brought up the shaft in nets.

Mine owners in Cornwall were even worse than those in the North East in providing adequate access to their mines. For the miners in the Cornish tin mines during the nineteenth century, the most debilitating aspect of their day's work was the appalling climb back to the surface at the end of an arduous shift. The first 'man engine' was installed at Tresavean Mine near Redruth in 1842. A man-engine was an assembly of wooden rods, with hand and foot holds cut in it, which extended down the shaft. Platforms were fixed in the shaft every 10 or 12ft. The rod was connected to a Cornish engine with a stroke similar to the platform spacing so that the miners could be raised or lowered by stepping on or off the rods and on to the platforms at each alternate strokes. Cornishmen often had long tram journeys into the mine. During summer weather they might take a long (maybe last) lingering glance at the sea before the tram entered the mine entrance.

At the beginning of the nineteenth century, children who were going to work in the mines were often brought to the pit on their fathers' backs. My father, shortly after starting work at the age of twelve in 1912, suffered from arthritis, and was carried to work on the back of his elder brother (and marra). Once underground in the 'hot' pit, his limbs warmed up and he was able to work, but on his journey home above ground, he again was carried!

At Australian colliery Stanford Merthyr, in 1922, four sinkers working for East Greta Mining Co. were still using a horse whim. It would seem that Azariah Thomas, the colliery manager, had still to introduce steam power.

[Janet Cumming.]

[Girl carrying Coals.]

[Load dropping on ladder while ascending.]

Winching children down shaft at a Yorkshire colliery in 1842. (Parliamentary Archives)

Left and above: Coal bearers at work at a Scottish colliery in 1842. (Parliamentary Archives)

Sinkers, Miners and Mining Engineers Working Away from Home

During his reign, Edward I sent miners from Minera in Denbighshire to Cornwall to help in the mining of tin as well as bringing miners from the Peak District of Derbyshire to work in the Flintshire lead mines.

Sinkers, engineers and viewers became willing to travel to neighbouring counties as new pits opened. At the latter end of 1711 Thomas Newcomen made proposals to drain the water of a colliery at Griff in Warwickshire, and here during the period 1702–29, there was a great demand for miners, who were paid slightly more than the most highly paid craftsmen (freemasons). Similarly, two Leicestershire mines, Measham and Swanningham, where the wages were high, recruited men from Shropshire, Derbyshire and Warwickshire. Their wages would have included lodging money and sometimes binding money. These variations in demand and wages continued to make miners move between counties during the next two centuries.

With Durham and Northumberland containing the world's greatest coalfields, it is not surprising that Scotland and Cumberland should be net importers of mining expertise from their neighbour. Writing to Sir John Clerk in 1703, a well-known coal master of Woolmet Pit in the Lothian area, commented:

> Most of the Sinkers or mynders [Miners] in my works are Newcastle men wch I bring from thence, finding them incomparable befor ours both for work, honnestie & Civilitie, I know no way so fitt for you to provyde y[our]selfe as sending to Newcastle.

Even as late as 1793 a Durham sinker was noted as putting down a new shaft at Strathaven in Lanarkshire. English viewers were still sometimes employed to give an opinion on particularly troublesome Scottish sinkings.

In the North East, miners dissatisfied with one colliery would get 'bound' to another colliery, so that shifting of belongings and family to a new home was often necessary. Thus thousands of families would be on the move annually with the means of transport ranging from big wagons to 'cuddy carts'. A common phrase was 'follow the wark' when a miner had to face the reality of looking for work. Due to the nature of sinking, pit sinkers were on the move much more than ordinary miners. Employers sent agents around the coalfields to recruit/steal workers from each other.

William Chapman, leasee of Wallsend collieries, was made bankrupt in 1792 after a sinking failure. He then became agent for Boulton & Watt supervising the erection of a colliery engine in Tyrone. Colliery prospectors were also required to move to exploit new coalfields. John Aynsley from Northumberland moved to the Potteries in the late eighteenth century and became the first chairman of Fenton Park Company, a mining company in north Staffordshire. By the end of that century this colliery was well established with a Boulton & Watt steam engine. A second engine was purchased in 1802/3 for £1,476 8s 0d. About this time, several of the more successful and wealthy potters decided to invest in the colliery business and ensure a supply of good coal. Josiah Spode II bought into Fenton Park about 1802 and paid off the outstanding debts.

The owners of Ness Colliery in Cheshire had a long association with the North East. Records show that in the early nineteenth century, Robert Johnson, a coal owner from Northumberland, was employed as coal viewer and principal agent at Ness in Cheshire. George Stephenson also kept a continuing interest in the Wirral collieries. In 1819 he send Newcastle civil engineer Joseph Cabry with his eighteen-year-old son Thomas to install a new winding engine, with the intention that they return to the North East to erect a large

Descent of a horse down a mine
shaft at Creuzot, France, in 1879.
(Science & Society Picture Library)

engine at Tyne Colliery. Later when Stephenson was involved in the Leeds to Liverpool
railway, he sourced stone from quarries near Storeton on the Wirral. He also carried out
early surveys for the Wirral railways.

Due to the long-established links between Cornwall and Wales, many Cornish sinkers
were employed in pits in South Wales. In 1827 near Swansea, when shaft sinking became
problematic due to flooding of shaft at 500 gallons per minute, Cornish shaft sinkers had to
be brought in to complete the shaft sinking.

The availability of labour was a critical factor in the way the owners treated their
workforce. When a pitch proved unrewarding, the Cornish miner would 'jack up',
confident he would find an alternative berth in other local mines. In the North East this
was rarely the case. In the 1840s there was a decline in the Somerset coalfields, and most
of their redundant miners went to South Wales, with a few moving directly to the North.
From the early 1800s there was a steady influx of Irishmen to the British coalfields which
increased rapidly following the Great Famine. A rock fall in a tunnel east of Rouen (France)
in the 1840s trapped a French labourer. Lancashire navvy rescuers sank a shaft at 10ft/hour,
something of a brisk mole's pace! It was clear that navvies with sinking experience were
finding work all over Europe.

It was not surprising to discover North East sinkers such as Francis Hardy, a master
sinker, in nearby Yorkshire. Also Lancashire, Staffordshire, Cheshire and South Wales
attracted many North East sinkers who were to settle in these counties. In the 1851 census,
Mary Pread, living in Redruth, Cornwall, was recorded as 'supported by husband who is

a miner in Cuba'. Both Cornish and Northumbrian mining engineers were to be found working in South America where minerals were being extracted. The main sinker families also worked in many European countries including Prussia. Australia was also attracting the British skills developed during the Industrial Revolution. Chappell Bros of Australia deployed their master shaft sinker Eric Anderson in 1928 at the new Elrington Colliery in New South Wales. Francis Ford, a carter born near Radstock Colliery in Somerset, first moved to Hetton-le-Hole in Durham where he worked as a sinker in the 1870s. He then moved to Australia where he was to also to continue working as a sinker. British sinkers were found in South Africa, which was to become the leader in sinking technology as it developed its diamond mines.

The following examples illustrate this movement of mining folk from the North East:

1851	Joseph Cabry, engineer, Durham	at Bishophills, Yorkshire
1861	William Maddison, Durham	viewer at Thornhill, Yorkshire
1861	William Potter, Northumberland	viewer at Monk Bretton, Yorkshire
1861	William Weatherburn	sinker at Pembleton, Lancashire
1861	Thomas Emmerson, Durham	master sinker at Staveley, Derbyshire
1861	Alfred S. Palmer, Northumberland	mining engineer at Radstock, Somerset
1861	John Baley, Durham	surveyor at Radstock, Somerset
1871	Thomas Emmerson, Durham	pit manager at Wales, Yorkshire
1871	Henry Cowey, Northumberland	surveyor at Tankersley, Yorkshire
1871	William Mason, Durham	master sinker at Aughton, Lancashire
1871	George Greenwell, Northumberland	mining engineer at Payton, Cheshire
1871	John Oliver, Durham	mining engineer at Bedworth, Warwickshire
1871	George Saint, Northumberland	mechanical engineer at Merthyr Tydfil
1881	Francis Hardy, Durham	master sinker at Brotton, Yorkshire
1881	Thomas Bell, Durham	pit sinker at Brotton, Yorkshire
1881	John Oxnard, Durham	mining engineer at West Ardsley, Yorkshire
1881	William Hunter, Durham	mining engineer at Monk Bretton, Yorkshire
1881	Edward Henderson, Northumberland	mining engineer at Coleford, Gloucester
1891	James Beacon, Durham	pit sinker at Sandal Magna, Yorkshire
1891	William Coulson, Durham	pit sinker at Sheffield, Yorkshire
1891	Matthew Mills, Durham	pit sinker at Rotherham, Yorkshire
1891	William Tate, Northumberland	mining engineer at West Ardsley, Yorkshire
1891	William Oliver, Northumberland	mining engineer at Huntwick, Yorkshire
1891	Robert Hunter	mining engineer at Leigh, Lancashire
1891	Matthew Clark, Durham	colliery sinker at Aberystruth
1891	John Briggs, Durham	pit sinker at Abertyboy, Monmouthshire
1891	Robert Williamson, Durham	mining engineer at Aughton, Lancashire
1901	James Mason, Durham	pit sinker at Hanley, Staffordshire
1901	William Bell, Durham	coal mine sinker at Kingswinford, Staffordshire
1901	Harry Mason, Durham	pit sinker at Lingdale, Yorkshire

Of course, Cornwall and Devon continued to export its sinkers and mining engineers:

| 1851 | Thomas Trewellin | sinker at Aberdare, Merthyr Tydfil |
| 1851 | Richard Kestell | sinker at Aberdare |

1868	William Dover	sinker Castle Eden, Durham
1871	William James	sinker at Merthyr Tydfil
1881	Henry Bealey	winding engineer at Llanwonno
1881	John Stephens	sinker at Brotton, Yorkshire
1881	Mark Luke	sinker at Droylsden, Yorkshire
1881	Mark Williams	sinker at Aberyshan, South Wales
1891	John Lasson	sinker at Neath, Glamorgan
1891	Richard Thomas	sinker at Abertyboy, Monmouthshire
1891	John Vivian	mine and civil engineer Preston Quarter, Cumberland

The Vivian mining family from Cornwall also established a mining company in South Wales at Morfa Colliery.

British master sinkers and mining engineers were commissioned to visit and work in overseas coalfields:

William Coulson	Prussia (Westphalia – Hibernia and Banrock collieries) and Austria
Robert Coulson	Borneo
Francis Hardy	Prussia (Westphalia)
Thomas Lee	Castrop Westphalia Prussia
Parkin Jeffcock	Moselle Saarbrück, Prussia
Andrew Laverick	Prussia
John George Weeks	Prussia
William Moses	Teintain, North China, America and Japan

In the period 1853–4, honorary members to the North of England Institute of Mining Engineers included Baron von Humboldt of Prussia and Mons De Vaux, Inspector General of Mines, Brussels.

Accommodation

In the North East, before the rapid exploitation of the coalfields, the land was overwhelmingly farmland and open landscape, with isolated cottages and a few quarries. From the 1830s country areas were rapidly covered by the lofty steam-engine chimneys sending volumes of smoke into the sky. The Vale of Houghton in County Durham had been desolate and barren until the shaft sinkers moved in, but by 1840 an entire township with a population of 2,000 had sprung up around the pithead stocks. With the establishment of the pit villages, generally in remote places, the first to be housed were the shaft sinkers, followed by the jerry builders who erected the long rows of one-up, one-down cottages, to house the incoming miners. Where the coal mines were located near the ports, 'squalid havens of free enterprise' sprang up; pitmen lived with fishermen, seamen and factory workers jammed in terraces of back-to-back houses with enclosed courts, often linked to main streets by covered passages.

With sinkers employed at the opening of a colliery, their accommodation consisted of small stone cottages built from the first stone raised from the pit shaft as it was being sunk. The line of cottages were often called Stone Row or Sinkers Row, for example at Leasingthorne near Bishop Auckland, Wingate and Haswell. The sinker rows were often much larger and commodious than any of the other colliery stone cottages, although still small to accommodate a large family. As progress was made towards the full operation of the pit, permanent colliery housing was built and eventually the sinker huts were replaced and the 'raws' were taken over by the colliery owners. Unfortunately the collier families often settled in the houses before they were thoroughly dried out, leading to diseases such as consumption. The carts which delivered coal to the front, then went round to the back to remove ash and waste and empty it on adjoining fields. Women were sometimes able to use the warm water from the engine houses for their washing; however, sanitation was sub-standard giving rise to high mortality.

If sinkers were required to return to working collieries, they often lodged with other established mining families. This was probably when they came back to sink further shafts or deepen or widen existing shafts. Not surprisingly these sinkers lived in the same streets. In 1841 at Thornley, the first houses were built in Church Street for shaft sinkers who were largely from Cornwall, and there were a few German sinkers. The miners living in the Forest of Dean, Somerset and Cannock Chase were countrymen. Some of them lived in colliers' rows near the pits, but even here the cluster of miners' cottages fitted unobtrusively into the rural scene. However, the habits of miners were not the same as sinkers!

By the end of the nineteenth century, sinkers' housing had not improved. When the first turf was cut in 1899 for the sinking of Easington Colliery, the sinkers lived in huts not much more than shacks on site, similar to the tin huts used by the railway navvies who had a similar lifestyle to the pit sinkers. Most of the railway tunnel gangs lived in 'shants' around the tunnel shafts, each gang allocated to a separate shant or hut made from sods. Two decades later the sinker huts provided in Kent were not much better, simply constructed of wood or tin. At Thorne Colliery, Yorkshire, in 1906 when work commenced, the first sinkers lived in a small village of wooden huts with corrugated iron roofs. This village was located next to a quarry, where clay and sand was extracted in order to make bricks to line the shafts.

Stone row of sinkers' cottages at Barrington Colliery, Northumberland, painting by James Mackenzie, c.1905. (Beamish Museum)

4

Colliery Disasters

As early as 1556, Agricola drew attention to the high death rate from lung disease amongst the metal miners in the Carpathian Mountains. No doubt there were many deaths for other reasons but these were not reported. In the centuries to follow, mine owners and workers worldwide too readily accepted that the dangers came with the job. With the introduction of more effective pumping engines it became possible to mine the lower seams, but working in deeper shafts involved many more hazards and affected many more underground workers.

Even for those pit workers who did not go underground, the shaft was akin to the devil's mouth, not to be treated lightly. Before the installation of self-acting fences which automatically closed when the cage descended, many surface workers fell down the shaft; a great number of these were female workers. In 1856, Thornley Colliery gave recommendations for improving the safety around the top of shafts to stop tubs from accidentally falling down; nevertheless ten years later seventeen-year-old Hannah Rees was 'dashed to pieces' at the bottom of the pit belonging to the Tredegar Iron Company.

Getting close to the edge of the shaft remained a fear for all colliery workers. A Durham boy recalled his first visit to the pit mouth: 'The onsetter shouted for me … a big strong feller, got me in his arms and he held me near the shaft, he frightened my life out …' In early times none but the most intrepid colliers would venture down the pit. Two men descended at a time in a sinking corve, with a rope around their bodies to attach them to a chain. They grubbed around and filled the corve with their hands as well as they could and then returned to the top with it. This trip constituted their shift for which they were paid 5s each with a 'pour boire' into the bargain. Sometimes in sinking a shaft, the men worked for days in complete darkness, rather than risk an explosion.

The last mention of women working underground in the North East was an account of a shaft accident. In 1772, a woman employed in 'putting at South Biddick [was] riding up one of the pits [when] the other hook in p … caught her cloathes'. The weight of the rope forced her out of the loop, and she fell to the bottom of the shaft. This showed that although women were never employed as sinkers, they nevertheless had to face the dangers of the shaft. The earliest pit fatalities were recorded at Gateshead in 1621; an explosion was mentioned in an entry in the register of St Mary's church: 'Richard Backus, burnt in pit.' In 1648 George Rutter was 'slaine in a pit', as was Michael Laurin in 1692. Again in Gateshead, the first large-scale mining disaster was reported in 1705 at Stoney Flat Colliery, where thirty miners including one woman were killed in an underground explosion. At Whitehaven in 1737, documents stated that 'fire-damp killed 22'. It appears the sinking of shafts was probably the most dangerous activity in coal mining.

Early disasters involving a small number of deaths were rarely recorded, and there were no inquests. Collectively, the number of these fatalities was considerable. The plight of the many that were injured or disabled was given little publicity, and they were forced 'upon the parish'. In nearly every case, the survivors of those killed had to seek asylum in the workhouse, or wander the country as beggars. In 1756, an angry Newcastle newspaper editor demanded that owners should fill in old workings [shafts] to prevent people falling in; however, the more serious dangers underground were rarely raised in the press. Eleven years later, the *Newcastle Journal* commented with an air of resignation that:

> … it certainly claims the attention of coal owners to make provision for the distressed widows and fatherless children occasioned by these mines, as the catastrophes, from foul air, become more common than ever; yet as we are requested to take no particular notice of these things, which, in fact, could have very little good tendency, we drop further mentioning of it.

A lucky escape occurred in 1763 when a young collier, John Boys, working near Lanchester in Durham was about to descend the shaft. He prepared to step into the loop, but because of the thick dense vapour around the mouth of the pit, he over-balanced and fell down the shaft to a depth of 42 fathoms (252ft). When other colliers were sent down in a corve to recover the body, they discovered to their amazement that he was still alive. It was thought that his remarkable escape was due to having fallen 'perpendicularly' without having been dashed and reverberated from side to side of the shaft. Thereafter, he was unable to walk without the aid of two sticks, and he had to resort to cobbling old shoes to avoid entering the workhouse.

Anthony Errington, one of the few miners to write an autobiography, had a number of narrow escapes. At the age of eleven years in 1789, he played games around the top of a shaft at Hollihill Pit near Felling in Durham. (I remember at a similar age, whilst living in Durham, throwing stones over the circular wall at the top of a disused shaft and waiting to hear the splash many seconds later.) Ten years later, as a waggonway-wright, he escaped death after he had hung perilously from a rope in the Ann Pit shaft near Newcastle. Hay was even thrown down to the bottom of the shaft in case he could not hold on, in the hope that it might help to break his fall. He was much admired by Mr Buddle, the viewer of Wallsend, when he heard of Errington's quick actions in helping to avert a number of mine disasters arising from the presence of explosive gases.

The large amount of horse traffic in colliery districts was a particular threat to small children. George Stephenson remembered one of his duties, as a young boy, was to look after his younger brothers and sisters, to ensure they were kept out of the way of the chauldron waggons, which were dragged by horses along the wooden tramroad in front of his home. The tramroads or waggonways led to the staithes located at the river bank or the dock wharfs. At the docks, a 'coal trimmer' was employed in the extremely heavy task of throwing coal from wagons into boats. Sometimes it could be done more easily with the use of chutes and gravity, but more often the coal simply piled up below the hatches and did not reach into the corners of the hold of the boat. Trimmers then had to get down into the hold and make sure that the coal reached the corners. Due to working in a confined space, coal dust was a serious health problem.

Miners would have always been aware of the dangers of 'bad' gases and they used various ways to get advance warning of these potential hazards. They were alert to the groans and moans of the mine and the creaking of pit-props which might indicate the imminent collapse of an underground roof. Caged canaries were used by colliery officials to signal the presence of deadly gases. Canaries were particularly sensitive to methane and carbon dioxide. As long as the canary kept singing the miner knew he was safe, but a dead canary meant an immediate evacuation. Sometimes other animals rather than birds were used. Even as gas detection technology improved, some mining companies in the twentieth century still relied on a 'canary in a coal mine'. As in Britain, the death toll of miners in America due to explosions was substantial. During the rescue in 1869 at Avondale in Pennsylvania, a small dog in a bag and a lighted lantern were lowered down the shaft to see if the air was foul enough to kill the dog or extinguish the light. The risks associated with the fireman's method of holding a rod with a burning candle attached to its end was gradually abandoned with the introduction of various safety lamps. In the old small pits there was a simple method of avoiding disastrous explosions – at the sign of gas the men left the pit.

John Haldane, the famous Scottish physiologist, investigated the reasons for many colliery disasters, in particular the nature of the toxic gases which killed miners after an explosion. Black-damp was an extinctive gas or mixture of carbon dioxide and nitrogen which was usually met with in old workings or in ill-ventilated parts of coal mines, whilst occasionally it issued from the goaf (exhausted workings) due to a fall in the atmospheric pressure. In shallow mines it could exude in large quantities from the coal face. Marsh-damp or methane was known by miners as fire-damp or light carburetted hydrogen. As the name suggests, it arose in marshy areas or stagnant pools from the decomposition of vegetable matter. It was found in many countries – in Derbyshire lead mines derived from the Yorkshire shales, in salt mines at Stassfurt in Prussia, at Wieliczka in Poland, and in Belgium. Other forms of explosive gas giving rise to the term 'fiery mines' were:

Choke-damp   carbon dioxide, often called carbonic acid; known as 'stythe' by miners. It escaped with fire-damp from the coal face.
After-damp   the atmosphere that remains after a colliery explosion composed of carbon dioxide, carbon monoxide, nitrogen and sometimes sulphur dioxide.

In the late nineteenth century experiments carried out for the Prussian Fire-Damp Commission revealed that the presence of coal dust was found to be critical in extending the effects of an explosion. Explosions in the shaft area were less likely since the shaft acted as a direct means of venting the gas.

Spedding invented his 'steel mill' in 1760. This machine was long used for providing a supposedly 'safe' light in coal mines. It appeared to have been only used in the north of England. By turning a handle, a steel-rimmed wheel was made to revolve rapidly against a piece of flint. It was normally worked by a boy who sat beside the pitman at work, and the continuous shower of sparks it emitted cast a feeble glimmer of light around. This contrivance, besides providing a very small illuminating power, was found to be dangerous; sometimes the sparks ignited the gas, with tragic results. At Wallsend Colliery, the shortcomings of use of the 'mill' were highlighted when at Pit B in 1786, three men were severely burnt. When the overman who was 'playing the mill' saw the gas igniting the sparks produced by the mill, he continued to work in the dark. The miners had their own ways to avoid the dangers of explosive gases. At nearby Walker Colliery, traditionally before a shift the men and boys would examine the weather, and if the wind was blowing from the south-east and threatening rain, they would return to their beds, even if this meant losing a day's wages.

The disaster at Jarrow Colliery in 1805, resulting in the death of many miners, leaving widows and orphans, had such an effect on Dr Trotter, a physician resident at Newcastle, that he prepared a pamphlet addressed to coal owners and agents entitled 'A Proposal for Destroying the Fire and Choak Damps of Coal Mines'. He suggested a scheme of depriving fire-damp of its explosive properties by a process of fumigation, to be performed by means of an oxygenated muriatic gas – a neutralising system. However, Dr Dewar of Manchester pointed out some dangerous deficiencies in this scheme and put forward alternate proposals using steam instead of fire for improving the ventilation. Within one year, these concerns generated other useful contributions such as the use of 'independent systems of ventilation'. This was followed by a new scheme by master borer James Ryan, proposing the use of chemical agents to destroy the noxious gases. Two years later Ryan was to apply his ventilation remedies to one of the most 'fiery' mines in south Staffordshire, and succeeded in clearing the mine from fire-damp in twenty days. This was effected by the rearrangement of the ventilating current, connecting air-heads or passages in the higher parts of the coal with the upcast instead of the downcast pit, thus allowing the drawing off of gas, instead of blowing in air as was the previous practice. This led to the general adoption of this 'top-head method' in the Midlands. However, the viewers in the North adhered to the diluting system of ventilation.

Due to 'creeps' in the strata near Wallsend, the normal ventilation became disrupted, and the air currents were loaded with inflammable gas to such a degree that it was necessary to discontinue ventilation by means of furnaces. Therefore, during the years 1807–10, viewer Buddle employed other forms of ventilation including the use of a steam ventilator which discharged steam into the upcast pit a few fathoms below the surface, which created a rarefied air current. Other methods applied were the use of a hot cylinder and the air-pump. John Buddle Jnr and Thomas Barnes went on to minimise the use of a furnace by subdividing the workings into independent systems. Only seven years after the disaster at Jarrow, the explosion at Felling Colliery in which ninety-two of his parishioners perished had such an impact on the Revd John Hodgson of Jarrow church that 'braving the displeasure of the affluent Brandlings' he published an account of this mining disaster, and formed the 'Society for Preventing Accidents in Coal Mines'. This society immediately approached the eminent scientist Sir Humphrey Davy, instructing him to explore a remedy at all possible speed.

A local man had already been working on this problem. Willie Woolhave, a plumber, glazier and parish clerk, carried a reputation as an inventor in the Newcastle area, having been involved in the early production of a lifeboat. He was said to have invented the first safety lamp in 1813. It was like a large parrot cage enclosing a glass lamp; the air was supplied by bellows worked by the collier's knees, using air from the lowest strata in the mine. At about the same time, another safety lamp was invented by Dr William Reid Clancy, an Irish physician who practised in Newcastle. For two years the Clancy lamp was recorded as being used at Herrington Mill pit in Durham. It became known as the 'Glennie' by many of the North East pitmen, and this name was later used for all types of flame safety lamps. (Dr Clancy was involved in fighting the cholera epidemic in 1832.)

Until 1815, it was not the custom in the North East to hold inquests on the victims of accidents in mines. Sir John Bayley had made strong representations at the Newcastle Assizes in 1814 on the scandal of omitting all inquiry into the circumstances under which hundreds of persons lost their lives.

In 1814 George Stephenson, then an engine-wright, was alerted by a workman that the deepest main seam of Killingworth was on fire. He rushed to the mine-head and ordered the enginemen to lower him down the shaft in a corve. Once down the mine he led a number of miners to build a brick fire-wall to exclude the atmospheric air. The fire was extinguished and this saved both the men and mine from disaster. Within months improved safety lamps were introduced by George Stephenson with his 'Geordie' lamp, and then Sir Humphrey Davy presenting his 'Davy' lamp after a trial of the lamp in the Hebburn Coal Workings. Before the use of these safety lamps, the hewers in the fiery collieries had to work almost in the dark. Despite the fact that the Geordie lamp had been developed locally, Buddle gave his view that the Davy lamp was the preferred safety lamp for miners to use in the 'fiery' collieries. Many miners were suspicious of Davy lamps, as they were of steam engines to raise the cages and pump the water out, and these changes were regarded as innovations designed to cheat the collier out of the 'profits and delights of existence'. Although Davy refused to take out a patent on his lamp, renouncing an income of £5,000 to £10,000 a year, it allowed mine owners to work deeper and more dangerous seams and thus increase their profits.

Technological improvements in mining and quarrying also encouraged owners to take risks with health and safety. Dust levels were another occupational hazard that increased with mechanisation. A Government Committee in 1835 reported that:

> The practice of placing wooden partitions or brattices in ventilating shafts, is to be reprobated; the slightest explosion may remove them, thus the whole system of ventilation is destroyed and no timely aid can be rendered to the temporarily surviving sufferers. Your committee have reason to believe that this opinion is generally adopted in coal-mining districts. To this point they attach an importance, inferior only to the provision of a sufficient number of upcast and downcast shafts. They consider the evidence justifies the suspicion that the foul and free air courses are frequently too near each other, the communications not adequately protected, and that the lengths of air coursings are excessive, giving opportunities for leakage, interruption, and contamination. and that the temporary nature of the stoppings – often boards imperfectly united, sometimes mere heaps of coal – and their frequent disarrangement, inevitably produce dangerous consequences.

At Eppleton Colliery, the inquest in 1836 concerning an underground explosion concluded that a young trapper had neglected his duty by leaving the ventilation door open long enough to allow the build up of foul air. Very young boys were given the task of operating the trapdoors used to regulate the passage of air through the mine, an important part of the ventilation system. These 'trappers' or door-keepers had to work longer hours than the adult miners, and understandably, frequently fell asleep as they sat in total darkness when their candles ran out. Sometimes they would be stirred out of their sleep by the tap of the viewer's stick, or if unlucky by the fist of the overman. Mr George Jobson, colliery viewer at Willington, Heaton and Burdon Main collieries near Newcastle, in supporting the continued use of young trappers, stated, 'The children stand in awe of deputies and superiors better than the older ones. Old men fall asleep and fit men would never stand degradation of doing a trapper's work.' Of course, dangerous working conditions were not confined to the mines and factories. The death toll during the construction of the Box Tunnel near Bath in 1836 was put as high as 100, and most of the casualties were workmen who had fallen down tunnel shafts. When Brunel was queried on the high level of deaths, he was to reply, 'They don't have to work for me, I don't ask them to, and so I'm not too bothered about them.' These were brutal times with little time for compassion!

Based on the findings of a Royal Commission in 1840, it was recommended to the Home Secretary, Sir George Gray, that a Government Officer of Mines be appointed for each district where explosive gas was present in collieries. In the mining districts of Northumberland and Durham he found complete readiness to co-operate with any scheme. In 1850 a Mines Act was passed indicating that the principle of state intervention in the interests of safety in mining was established. This led in 1862 to the appointment of six safety officers, too late for the Hartley disaster and even then a very small number to inspect the whole of the British coal industry. Such a small team could only cover major incidences such as explosions causing large loss of life, for example the Blantyre explosion in 1877 when 207 died. By the end of the century a further thirty inspectors were appointed, and by the time of the revised Act of 1911 the number of inspectors had risen to ninety.

There continued to be general unrest in the collieries country-wide regarding working conditions. As well as the men's complaints about fines and long hours for the boys, the pitmen considered that with the introduction of the Davy safety lamp, there had been an increase in temperature in the mines due to the neglect of ventilation, but 'men's complaints' were ignored by the coal owners. Following the Haswell disaster in 1841, and the inadequacy of the Davy lamp, the South Shields Committee proposals received support from Dr J. Murray for the need of more shafts in each colliery. James Mather, a North East miners' leader, wrote to the Prime Minister Sir Robert Peel demanding safety legislation. As a result the Government appointed Professors Lyell (founder of modern geology) and Faraday to attend the remaining Haswell hearings, and they produced the first scientific report on the causes of colliery disasters. Six years earlier Buddle had argued at a Select Committee that the cost of sinking additional shafts was cost prohibitive. The continuing colliery disasters started to have some impact on mine owners.

Rather than explosive gases, the main causes of shaft injury and death continued to be due to sudden inflows of water and falls of rock and other materials from above; however, there were other dangers. As the mines got deeper, temperatures increased. In the case of the Wearmouth Colliery, with its depth more than four times the height of St Paul's Cathedral, the average temperature ranged from 78 to 80 degrees Fahrenheit. The colliery surgeon described a type of boil affecting the miners caused by the irritating nature of the water in the pit. This especially affected new workmen for the first two or three months down the pit. A young mining engineer, John Elliott, gave evidence to House of Commons Committee on conditions at Wearmouth Colliery, sunk to a depth of 265 fathoms. He explained that it took two and three minutes respectively going down and coming up the shaft. The tub for drawing the coals, and also men and boys, was 7ft high. It held 105 Newcastle pecks of coal, weighing 30 hundredweights, and the cage rope itself weighed 5 tons. Ironically, in 1849 the same John Elliott, an under-viewer of twenty years of age, was killed accidentally falling down the shaft at Wearmouth.

It is difficult to determine whether any children were involved in sinking work, but it was true that some elderly sinkers were still working until injury or death intervened. Due to years of resistance by owners to improvements in miners' working conditions, the miners were suspicious of any changes proposed by their employers. The well-known mining engineer Nicholas Wood considered it unfortunate that the children of hewers and pit families generally were not in sufficient numbers to meet the demand, and the boys of mechanics and labourers in outlying areas were being sought to fill the gap. He noted, 'Pitmen with families are much sought after. As the pits go deeper we will need more boys.' Matthias Dunn, viewer and part owner of St Lawrence Main, a colliery just outside of Newcastle, was of a similar opinion, and did not consider that legislative interference was necessary. Nevertheless, his colliery was the first to introduce cages in shaft-winding, probably since it increased efficiency. Dunn was later to be appointed a Government Inspector of Mines. Even though the 1842 Mines Act was passed, it was a dead letter since no mining inspectors were appointed.

The first important court case brought by the workmen, was the outcome of a strike at Wingate in 1843 over the introduction of a wire rope to the shaft winding engine. The miners were suspicious of anything new that management wanted, assuming that there was an ulterior motive. Despite the many deaths caused by rope breakages, almost as soon as new type of rope was placed in the shaft the men refused to descend the shaft, not wishing to entrust their lives to anything but the old proven hempen cable. The owners were keen to use the wire rope due to its lighter weight, which permitted increased winding speeds thus allowing more efficient working in deeper mines. Eventually the owners got their way, or persuaded their workmen 'to see the light'.

In 1844 a major disaster took place at Haswell Colliery. Many of the victims lived at Sinkers Row such as the Briggs family, John aged twenty-five and son James aged ten. To commemorate those who had died, a poem was written by George Werth and translated from German by Laura Lafargue, daughter of Karl Marx:

> The hundred men of Haswell,
> They all died on the same day;
> They all died in the same hour;
> They all went the same way.
> And when they were all buried,
> Came a hundred women, lo.
> A hundred women of Haswell,
> It was a sight of owe!
> With all their children came they,
> With daughter and with son:
> 'Now thou rich man of Haswell
> Her wage to anyone!'
> By that rich man of Haswell
> Not long were they denied:
> A full week's wages he paid them
> For every man who died.
> And when the wage was given,
> His chest fast locked up he;
> The iron lock clicked sharply,
> The women wept bitterly.

On the banks of the Tyne and Wear, the mining engineers were taught the latest science and the practical skills were gained by the best and enlightened training from management. In contrast in the Staffordshire coal district, operations were largely carried out by rule of thumb, however its pits were on a smaller scale compared to those of the North East, particularly regarding depth. A Government mines inspector, in 1854, informed a Select Committee that in south Staffordshire: 'There was no system' and 'the pits are turned over as soon as they are sunk, to the butties'. The butty (or viewer as he would be termed in the North), or the ground bailiff, visits the colliery once week or once a fortnight, by himself, or with his deputy, measures the dead work, and enters it in the book. 'The management is entirely in the hands of the butties'. This contrasted with the North East where a daily inspection was carried out by the deputy overman and then followed up by the overman and one of his deputies with regard to the gas in the mine. The Staffordshire pits were ventilated solely by the means of the vast number of shafts by which the whole coalfield is honeycombed. Seldom was any thought given to any kind of artificial means of creating a current of air. The workers could not breathe at any considerable distance from the shafts; the consequence of the whole system is that the coal was worked in the slowest, most dangerous, and least economical fashion.

In 1858 at St Helens Petty Sessions in Lancashire, a drawer, Thomas Cunliffe, was charged with a breach of the Special Rule No.9, smoking in the Rushy Park pit. Lamps were exclusively used in the mine and he had unlocked his own lamp. He was sentenced to six weeks' imprisonment at Kirkdale Prison. In contrast Ralph Hunt, the undermanager at a St Helens colliery, was charged with smoking in the pit but there was insufficient evidence and the case was dismissed. Although this might suggest that there was one rule for the workers and another for management, at least it illustrated that safety was now a serious issue.

The New Hartley Colliery disaster in 1862 was to have a seminal impact on the mining industry, and the increased pressure on the Government to act could not be dismissed out of hand. Perhaps this was to affect the mood nine days later when a court case was reported by the *Derbyshire Times* on 25 January 1862 which showed the power of the Government Inspector John Hedley, born in Hartley, in dealing with the failure of the owner of Boythorpe Colliery to abide by the colliery rules. Hedley reported on the three main deficiencies:

1 – it was not acceptable for the general colliery rules to be posted inside the engine house since 'the colliers are prohibited from going into it [engine house], except by express permission'.
2 – the shaft had not been cased, with loose pieces on the shaft side which 'if they had fallen upon the heads of those at the bottom of the shaft, it would have done them great injury'.
3 – 'not providing proper means of communicating from the bottom to top of the shaft'.

Although the defendant was charged on all three counts, and the sums involved were small, the owner's counsel was clearly unhappy that the indictments had even been presented 'saying that he could conceive nothing more immoral or more disgraceful to insinuate a want of public duty towards a gentleman who had to fulfil a most responsible and a most unpleasant duty'.

Only seven years after the Hartley disaster, a mine disaster in America at Avondale, Pennsylvania, provoked a similar public outcry. A wooden breaker constructed over the shaft opening to the underground workings caught fire. The shaft was the sole means of exit from the mine; consequently the men working underground were trapped and died of suffocation. In America this disaster led to new mining regulations being enacted, mandating the requirement of double-shaft mines, and the prohibition against collieries being built directly over the mine shaft. At this time in the USA, public indignation resulted in the formation of a protest group known as the Molly Maguires, who raised havoc in the anthracite fields, later being charged with the deaths of a large number of mine bosses. A similar fire event took place in Britain, when a wooden cabin erected on the colliery hempstead near St Helens Lancashire caught fire in 1867. The smoke and fumes were drawn down the downcast shaft and into the intake roadways. No Fenian reaction resulted in this case!

Mine owners could have exerted tremendous influence on improving health and safety in collieries. Most owners fought tooth and nail against attempts to introduce legislation which would have forced them to implement extra safety measures such as the sinking of extra ventilation shafts to increase the flow of air and reduce the build up of toxic and explosive gases. In August 1872 the Coal Mines Regulation Act came into being which included inspection of the operation of cages used in colliery shafts. In 1882 following an explosion at Whitehaven Colliery in 1879, mining engineers W.N. and J.B. Atkinson published their book *Explosions in Coal Mines* which became a 'bible' for safety in the coal industry.

At this time a report on deaths in coal mines presented to the British Parliament gave a list of ways miners could be killed:

1 – falling down a mine shaft on the way down to the coal face.
2 – falling out of the 'bucket' bringing you up after a shift.
3 – being hit by a fall of dug coal falling down a mine shaft as it was lifted up.
4 – drowning in the mine.
5 – crushed to death.
6 – killed by explosions.
7 – suffocation by poisonous gas.
8 – being run over by a tram carrying out dug coal in the mine itself.

A Royal Commission on Accidents in Mines was set up in 1879, and, after a long delay, reported in 1886. The chairman was Sir Warrington Smyth and among its members was Sir George Elliott, Professor John Tyndal, and representing the miners, Thomas Burt. Sir Smyth was a mining engineer and professor of mining. Sir Elliott had started work as a trapper boy at the age of nine and rose to become a colliery owner and Member of Parliament. He considered that a proportion of the industry's profits should be paid into a fund for retired miners. John Tyndal was professor of physics and successor to Michael Faraday, and John Burt had been elected a radical Labour MP for Morpeth in 1874. The Commission's scope was 'to inquire into accidents in mines and the possible means of preventing their occurrences or limiting their disastrous consequences'. They covered all aspects of mining practice and made a major contribution to mining safety by bringing to light the danger of explosion from coal dust. Unfortunately, pneumoconiosis caused by dust on the chest still became a common death sentence.

The details of the main categories of shaft disasters are given below. They describe the carnage of shaft and mine workers over two centuries and reflect the misery and hardship of the large number of families affected, in most cases by the unnecessary loss of life:

Explosions

Without doubt, underground explosions caused the worst disasters with enormous loss of life. As mining operations started to involve larger-sized collieries, the significance of greater numbers getting killed was becoming difficult to ignore. Although the major explosions did not occur in the shaft, this main access route played a vital part in the escape and rescue efforts.

At Burdon Main in 1835, an explosion was caused by a candle used by a deputy carrying out his safety check! In 1849, a Bill was introduced by Mr Duncombe, a Chartist MP, to prevent the use of gunpowder and candles in mines, but the Bill was not passed, causing much public anger. However, in 1858, a miner just escaped a prison sentence but was charged £1 19s 0d by the colliery manager due to using candles, being in breach of the Special Rules of Hollingwood Pit, Derbyshire.

In 1864, the son of a master sinker and friend of George Stephenson, was severely burnt after an explosion at Seaton Colliery. At Bigge's Pit, Willington, sinkers employed at some distance in the east headway, were made aware of their danger by the rush of dust and wind; fortunately for them after the explosion of fire-damp had spent, they were able to reach the shaft uninjured. The use of candles with naked flames was the main cause of the explosions, but other areas of risk included furnace fires used for ventilation, and the lighting of fuses during the operation of blasting.

Boiler explosions above ground also led to fatalities. The explosion at Spennymoor, Merrington in 1857 was apparently due to poor maintenance of a boiler. The brakesman had attempted to stop a leak using a mixture of meal and horse dung, and the colliery engineer took the wrong decision to keep on working rather than 'lay off' the boiler. In 1873 a boiler explosion cost the life of the engineman at Seghill due to deterioration of the boiler plates not readily detected by ordinary inspection.

Cradle Ropes and Chains

There were many reasons for deaths resulting from workmen falling down the shaft. One principal factor was the breakage of the rope or chain supporting the sinkers' cradle or scaffold. However, for most collieries after 1842, chains were outlawed and flat ropes were used for shaft winding; no wire ropes were used other on the railroads.

At Gateshead in 1834, a corf became unhooked resulting in the deaths of four workers. In 1840 at Springwell, the chain gave way on the cage causing all the occupants to fall down the shaft. Seven years later at Percy Main, a horse-gin was bringing up two men who had been examining the pumping gear used for drawing the water from the mine. The winding rope broke having got entangled in the scaffolding.

Flooding

The flooding of mines mainly took place by penetration of water into the working area due to flows through permeable strata and natural fractures. This was of particular relevance in the sinking of shafts where systems of lining of the shaft walls were developed to minimise this problem. However, the larger losses of life were often due to the unexpected inrush of water which prevented remedial measures being taken quickly, and therefore delaying the supply and setting up of pumps. Mining operations close to old workings carried the risk of releasing large quantities of water.

At Heaton Main Colliery in 1815, water broke in from old workings despite recent work to drain water from old shafts by driving a drift. Seventy-five workers were drowned; the immediate use of pumps to bring down the water level proved unsuccessful. It took many months to pump the water down sufficiently to recover the bodies.

A serious incident occurred at the Union and Isabella pits in Workington in 1837 when they were flooded by inflows from the Irish Sea. Twenty-five men and two boys were killed and 300 more laid off. At this time there was work available only two days in the week. Disasters had an immediate effect on employment.

At Tynewydd Colliery in 1877, a vast inundation of water burst in from an adjoining pit due to an inadequate barrier separating the pits. In order to rescue the trapped miners, the overman decided to sink a shaft to the heading below, but after the sinking, discovered the bodies of a miner and a boy drowned by the rising water. The engineers then instructed a rescue attempt at a higher level, and after pumping to lower the water level, a tunnel was driven to a particular stall. However, after trial boring, an explosive mixture of gas filled the tunnel, identified by the flames in the Clanny lamps flaring dangerously. Heavy temporary trapdoors were erected behind the rescuers excavating through to the trapped miners to confine any potential explosion. This rescue of four men and a boy who were close to starvation was considered one of the most exceptional rescues in coal-mining history.

At Devon Colliery, Clackmannanshire, in 1897, the County Constabulary descended the shaft to record evidence on the state of the flooded mine. In England, the investigator would have been an Inspector (HM) of Mines. This showed at long last that health and safety negligence in mines was being treated as a criminal matter.

Shaft Lining Failures and Falls from Above

Falls of material either from the sides of shafts or from above led to many accidents, caused by chance happenings, carelessness or poor maintenance. The dangers in lining the shafts occurred when encountering broken ground, where there was a 'fault' in the dipping strata, so that instead of solid rock there was crushed material, of the nature of clay and shale. Leaving a section of this ground unlined for any length of time placed the sinkers at great risk. The sinkers described this risky ground as 'a little bit tender'. The sinkers always kept the kibble on 'tight rope' attached to the scaffold. During the process of lining, it was especially dangerous to loosen the crib in order to complete the bricking up. In variable ground there was always the potential of a sudden inrush of loose material often followed by water flows.

At Harrington Pit near Whitehaven in 1881, a contractor was carrying out work in the air staple shaft. It appeared that a stone fell off a bunton, where it had lodged after blasting operations, falling on the sinker below. It was judged that after blasting, an inspection should have been carried out. Twenty years later, a collapse of the sides of a shaft caused the deaths of four sinkers at Croft Pit, Whitehaven. At Lady Durham Colliery in 1905, stones which were being loaded into a kibble from an intermediate seam tipped over the edge of the kibble and killed a sinker at the bottom of shaft. Another sinker was killed in a similar incident at Lumley in 1907 when old shaft stones, which were being raised, got caught accidentally, and one of the stones fell down the shaft, hitting the sinker on the head.

Lack of Experience, Mischief, Ill Health or Bad Luck

There have always been deaths caused by human error, ill health or due to factors outside of normal behaviour, but the shaft was rarely forgiving.

At Butterknowle in 1848, a banksman wandered to the pit, out of hours, after a visit to the beer house, and fell down the shaft. In 1887, Robert Bell, a railway fireman, committed suicide by throwing himself down the shaft. Four years earlier at East Howle, a boy trapper of thirteen years returning to bank in the cage was decapitated as he inadvertently put his head out, and came in contact with the main or kep bunton. In 1919 at Friars Goose, a miner accidentally fell down an old disused pumping shaft. It appeared that he was going to destroy one of his dogs. His body had to be recovered by ropes and grappling irons. The Kilsby Tunnel, designed by Robert Stephenson in 1833, employed 1,250 navvies and took two years to complete. During construction, two or three workmen were killed trying to jump one after another in a game of 'Follow My Leader' across the mouths of the shafts.

Foreman-sinker Thomas Emmerson (brother of a master sinker of New Hartley) was working at the Seymour Pit in Derbyshire, named after the mining engineer from Durham, Martyn Seymour. The accident took place on the Speedwell branch line belonging to Richard Barrow Esq., owner of the Staveley Works. Emmerson was attempting to get upon the locomotive engine when in motion. It appeared that his foot slipped on the step, and he was jerked down; one of his legs got caught across the rails and was passed over by several wagons. His leg was all but severed. Fortunately the surgeon and doctor were quickly available and an amputation carried out successfully. Despite this major injury, Emmerson went on to be master sinker at Staveley as well as a colliery manager in Yorkshire.

Master sinker George Sousby was descending a shaft to collect a charge of explosive in 1893 at Montague Colliery. He was riding with one leg in and one leg out of the kibble and either through carelessness or due to sudden faintness fell out of kibble down the shaft. Another master sinker, John Jackson, was on his way home for breakfast in 1903 when he suddenly became ill and died on the roadside; death was thought to be due to a heart attack. It appears that no inquest was considered necessary, and the fatality was probably classed as 'death due to natural causes'. It is not surprising that Jackson, still working in such a demanding job at the age of seventy-three, should succumb to his exertions.

Maintenance of Machinery

Some accidents were due to poor maintenance and inadequate setting up of colliery equipment such as winding gear which often led to fatalities, however, it is unclear in some cases which party held the principal responsibility.

In 1843 two engine-wrights at Wearmouth Colliery were carrying out maintenance work on pumps in the shaft. One of the men was John Coxon, from a well-known sinker family. They were working in the back shaft separated from the principal part of the shaft by means of a brattice, and were descending the shaft in a sling or loop in which a piece of timber was fixed as a seat. They had not gone far when one of the pump spears broke off, hitting the men with great force, and precipitated them down the shaft. They were either smashed to pieces on the wooden crossings which supported the pumps, or fell to the sump at the bottom of shaft. In the recovery of the bodies, one body was found in the 'cistern hole' 60 fathoms from the pit bottom.

At Dryburn Colliery in 1856, the horse-gin was positioned higher than the pit pulleys. The rope fell off the gin-roll to the axle, jerking the rope which broke, precipitating a man down the shaft. In 1857 at Kelloe Winning, a sinker was ascending astride the winding rope in order to attach part of the pumping equipment called the 'strum'. The strum got jammed and became detached from the rope, sending the sinker to the shaft bottom. At Elemore Colliery in 1857 a horrific accident took place when an old pitman was ascending a shaft by means of chain loops (which had discontinued in most other collieries). This pitman got entangled with a descending loop which pulled the poor man's head from his body.

Cradle or Platform Instability

The upsetting of the cradle was often caused by the difficulty of working by sinkers within a confined space. In 1856, the Mine Inspector's report concluded that three sinkers at Wardley Colliery had been killed by mismanagement of the cradle. At Rainton Meadows Pit in 1881, three sinkers were working on repairs to the pit scaffold at the bottom of the pumping shaft. As the chargeman, an engineer, was passing the cross bunton used for staying the pumps, it seemed that he had lost balance and fell from the 'loop' on the 'jack rope'. Two years later at Seaton Delaval, a rope supporting the cage suddenly broke sending the cage and the shaftman to the bottom of the shaft with another worker falling across the bunton. The two men were allowed to access the shaft by standing on top of the cage. It was considered that the colliery engineer should have examined the cage fittings himself.

At Dawdon in 1907, the cradle snagged on the shaft side, but with only three chains attached, the cradle overbalanced, tipping three sinkers down the shaft. Just earlier at the same colliery, a falling electrical cable caught the men in a kibble killing two sinkers and injuring the third. At Barrington in 1905, a cradle hung in the shaft at a crib bed 5 fathoms from shaft bottom, and was suspended on a crab rope. The kibble hung on a rope in the centre of the shaft, and the cradle rope gradually approached the kibble in descending and eventually pressed against the kibble. The sinkers had steadied the kibble on the wrong side of the cradle rope, and the kibble hit against the buntons, causing the sinker to be knocked out of the cradle.

Communication

Poor communication was often to blame for shaft accidents, particularly between master sinker or chargeman and the engineman. A rapping rope was used to transmit signals between shaft workers and the surface. It was also common practice for the end of a shift to be communicated by the order 'loose, loose' being shouted down the shaft, and then this was relayed to the furthest extremities of the pit.

At Wingate Grange in 1903, a pit was being sunk below the working shaft by means of a kibble under the ordinary coal cages. When the master sinker signalled to the engineman to lower the kibble, he found that the rope was becoming slack. Having recognised the danger they were in, he attempted to get the sinkers out of the kibble, but failed, and when the rope jerked back the sinkers fell to the shaft bottom. The accident was caused by one side of the shackle catching on the scaffold girder flange. After this incident, the manager revised the routing of the rope through the scaffold to avoid this type of risk in future, unlike the previous lack of response by management.

During the period between 1876 and 1900 there was a rapid reduction in accidents. However, the introduction of machinery caused accidents from contact with the machines, greater noise levels, and making it harder to detect the faint cracking, signalling an impending roof fall. After one of the worst mine disasters at Gresford in 1934, the Royal Commission recommended that for large collieries a mining engineer superior to the manager should be appointed, of strong character to resist commercial pressures. The inquiry into Gresford also highlighted deficiencies in ventilation, egress and coal dust.

Accident Relief

In the North East an injured miner was allowed to continue to live in his colliery house and receive his allowance of free coal. Also he was paid an accident allowance, known as 'smart money'. In 1880 this amounted to 5*s* per week, and half this amount for boys. In other counties the extent of these benefits was less generous.

Towards the latter part of the nineteenth century the unions were beginning to succeed with compensation claims for injuries at work. In 1887 the *Prescot Reporter* informed its readers of an accident to a sinker, Thomas Kane of St Helens, who had claimed £30 from contractor Isaiah Pilling, on account of personal injury he received when sinking a shaft at Bold Colliery (Lancashire) in 1886. This County Court case was settled with the plaintive receiving twenty guineas and each party paying their own costs. (Strangely, my great-grandfather, a master sinker, died at Boldon Colliery in 1886 from shaft injuries.)

My great-grandfather was also involved in the sinking of colliery shafts at Wingate. During the disaster of 1906 at Wingate Grange, his son James led the rescue of the trapped miners. James had been a master sinker at the Deep Pit at Hanley in Staffordshire in 1901. His father had passed on to him the knowledge of the access passages at Wingate, and he was able to save a large number of miners from certain death using an escape tunnel. During the rescue he was badly gassed, and was unable to work again. The colliery owner, in recognition of his bravery, allowed him to continue to live in the colliery house, but ten years later this 'favour' was rescinded, and he was forced to leave his home.

In the 1880 explosion at Seaham Colliery, an empty water bottle was found with one of the pitmen who died entombed. The message, scratched on the tin with a nail, eloquently expressed the emotions of one miner shortly before his death:

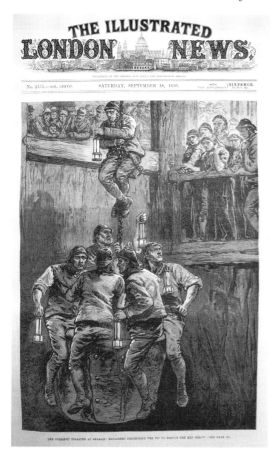

Left: Rescuers in a kibble at the top of the shaft during the Seaham Colliery disaster, Durham, published in *The Illustrated London News*, 15 September 1880. (John Weedy Photos)

Right: Miner's last message to his wife, 1880 – this empty water bottle was found with one of the pitmen who died in the Seaham Colliery disaster in 1880. (*Life and Tradition in Northumberland and Durham*, Frank Atkinson, published by J.M. Dent & Sons)

Dear Margaret,
There are 40 of us altogether at 7am. Some was singing hymns, but my thoughts was on my little Michael that him and I would meet in heaven at the same time. Oh Dear Wife, God save you and the children, and pray for me ... Dear wife Farewell. My last thoughts are about you and the children. Be sure and learn the children to pray for me. Oh what an awful position we are in.
Michael Smith, 54 Henry Street.

Despite the continuing toll of sinkers killed resulting from material falling from above, little was done to provide them with protection. In 1944 Will Lawther JP, President of the Mineworkers' Federation of Great Britain, summarised the grim record of the coal industry as 'blood on coal' based on the fact that 'no other industry has such a record of hardship, poverty, endurance, terrorism, disaster, unemployment, and of constant menace of death'.

The ultimate recognition of a mining engineer's competence and respect on all sides of the industry was his appointment as HM Inspector of Mines. It is pertinent that just prior to the rescue of the Chilean miners in 2010, there had only been three inspectors to check on health and safety for 884 mines in that area. Miners accept many dangers but resent unnecessary risks due to false economy and managerial short-sightedness. Most accidents were not 'inevitable'.

New Hartley Colliery Disaster

The disaster in 1862 at New Hartley Colliery, a small community near Blyth in south-east Northumberland, was probably to have one of the most important impacts on coal-mining

Above: William Coulson with sinkers, photograph by W. & D. Downey 1862. (Royal Collection)

Right: The obverse and reverse views of the Hartley Rescue gold medal. (North of England Institute of Mining and Mechanical Engineers: NEIMME)

Far right: A twelve-year-old tally boy. (NEIMME)

110 THE ILLUSTRATED LONDON NEWS [Feb. 1, 1862

THE FATAL ACCIDENT AT NEW HARTLEY COLLIERY.

MR. ROBERT TURNBULL, PITMAN'S DEPUTY. MR. COULSON, SUPERINTENDENT OF THE EXPLORING PARTY. MR. THOMAS WATSON, PITMAN.

MR. DAVID WILKINSON, MASTER SINKER. MR. EMMERSON, MASTER SINKER. MR. WILLIAM SHIELDS, MASTER SINKER.

VIEW OF HARTLEY COLLIERY FROM THE RAILWAY, TAKEN SHORTLY AFTER THE CATASTROPHE.—SEE PAGE 121.

Main rescuers, sinkers and pitmen published in *The Illustrated London News* (NEIMME):

1 Robert Turnbull, Pitman's Deputy.
2 William Coulson Snr.
3 Thomas Watson, Pitman.
4 David Wilkinson, Master Sinker.
5 George Emmerson, Master Sinker.
6 William Shields, Master Sinker.
7 View of New Hartley Colliery.

safety. For many years this mine located close to the sea had been subject to flooding, and as early as 1760 the mine had employed atmospheric and then steam pumping engines to control the frequent inundations.

The original Hartley Colliery (Whin Pit) did not have a happy history; Mr Curry the viewer died in an explosion in 1761. In 1834, four years after the Mill Pit opened, four miners were killed whilst descending the shaft when the rope broke. The new pit, Hester Pit, opened in 1845, and fifteen years later a miner fell to his death down the shaft. Also in 1860 a hewer met his death by falling down the shaft due to a misunderstanding between the waiter-on at the top and the workers at the bottom of the shaft. The cage had struck the stoppers unexpectedly and the hewer was thrown out of the cage. The colliery manager then announced that 'the waiter-on being an old man was to be replaced by one more alive to the important office'!

New Hartley pit was 100 fathoms deep, 15ft across and lined with timber and masonry. The colliery had three coal seams: first 'High Main', which was completely worked out by the time of the disaster, next 'Yard' seam and lastly, 30 fathoms below, 'Low Main', which

THE FATAL ACCIDENT AT NEW HARTLEY COLLIERY: REMOVAL OF THE COFFINS CONTAINING THE BODIES.

Removal of the coffin. (NEIMME)

was liable to flooding. Ironically Hartley was the first North East colliery to use a patented steam-powered system for movement of men and materials up and down the shaft. New Hartley Colliery, owned by Messrs Carr Brothers, was sunk by contractor Messrs Jobling, Carr & Co. The operators of this new pit had engaged William Coulson, the master sinker of Durham City, to supervise the shaft sinking of this ill-fated colliery, and therefore it was not surprising that in the event, he was called on to lead the rescue operation. As with the older pit, New Hartley was plagued by flooding. On one occasion all the workforce had to be withdrawn as the water rose 420ft up the shaft. A 300hp beam engine was installed to pump the water in three stages from the bottom of the 600ft shaft. The 42-ton cast-iron beam of the pump pivoted above the mouth of the shaft and was capable of raising over 1,500 gallons per minute. This made it by far the largest capacity pumping engine in the north of England.

The natural instinct felt by miners was the need for a second shaft as an escape route. Up to the mid-nineteenth century, mines generally had a single shaft divided into two halves by a timber partition called a brattice, separating access from ventilation and services. This was the arrangement used at New Hartley. Even though a Select Committee in 1835 and a South Shields Committee in 1841 had recommended that all mines should have at least two shafts, little was done by the colliery owners, always giving cost as the main objection. However, some colliery owners had been able to provide two shafts. Even after the Hartley disaster, mine workers had to wait until 1872 when the Coal Mines Regulation Act required the legal provision of two shafts for all collieries.

The fateful day fell on Thursday 16 January at New Hartley when the main beam to the pumping and winding engine broke and lodged in the shaft entombing most of those working underground. As half of the beam weighing about 21 tons thundered down the shaft, some sections of the shaft lining were dislodged, which made the rescue efforts very difficult, and blocked all attempts to reach those underground for several vital days. At the time the beam broke, a cage containing eight men was ascending the shaft. Six of the

Miners' family in cottage reading Queen Victoria's letter. (NEIMME)

Funeral cortege. (NEIMME)

occupants were killed instantly and the others miraculously escaped. Due to pump failure, this allowed the mine to flood, and the supply of fresh air was cut off with 204 men and boys succumbing to the deadly effects of 'stythe' or choke-damp. About thirty horses and ponies also died, probably drowning in the flood.

By Saturday afternoon the work of clearing the shaft was being carried out by relays of men at 2-hour intervals. Clearance work continued without break, and on Monday night the sinkers had even been too busy to 'jowl' down to the entrapped miners below. At one stage the master sinker thought there were signs that the men were still alive but by Wednesday all hope had evaporated. During the relentless rescue efforts in the bitter January weather, miners' families, doctors and mining engineers gathered around the top of the shaft, and this continued for a full week, day and night, awaiting news of their loved ones. Under master sinker William Coulson, in charge of the rescue, were other master sinkers, George Emmerson (his chief assistant), William Shields and David Wilkinson. The colliery's viewer, Joseph Humble, played a vital role keeping communications going between rescuers and the anxious families. The sinkers involved in the rescue had to be lowered to their work at the end of a rope in which a loop had been made, secured around their bodies. The two chief dangers they had to face were the continual falling of the sides of the shaft and the presence of the noxious gases. Shaft sinkers always knew that they could be called on to carry out rescue work, and these acts of heroism were rarely known or recognised by the general public.

The furnace smoke of the pit had been generating, amongst other fumes, the deadly carbon monoxide. Owing to the shaft blockage this gas had been accumulating in the Yard Coal in the immediate neighbourhood of the shaft, and it was this gas which had nearly destroyed the lives of some of the sinkers that Tuesday morning. Following a meeting of viewers, a decision was made to form an 'upcast' and 'downcast' from the main seam down to the Yard seam by means of a cloth brattice, and this took about one day to accomplish. A separate meeting was held on 22 March 1862 at Newcastle with the sole purpose of reporting on the condition of shafts for North East collieries. The Hartley disaster finally pushed the Government to take a major step regarding the improved provision of pit shaft access.

The night watchers. (NEIMME)

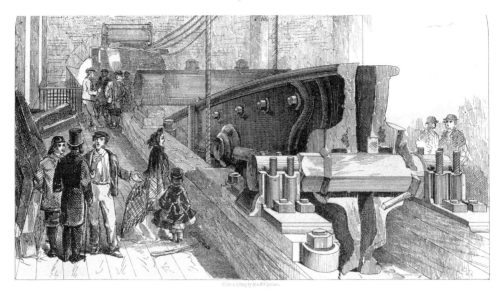

The broken beam. (NEIMME)

At one stage during the rescue at Hartley, a group of relatives anguishing over the doom facing their loved ones turned their bitterness against the master sinker leading the rescue. One man, shouting 'Shoot Coulson', threatened to take matters into his own hands, and another man, who had lost four sons in this tragedy, attempted to throw himself down the shaft. There was also the claim that ordinary pitmen were being excluded from the pit-mouth area whereas 'respectably dressed' people were being treated in a different manner. Mr Coulson addressed the crowd, and insisted that if this behaviour continued he would withdraw his sinkers, feeling deeply hurt by their reaction. The crowd probably saw Coulson as someone in collusion with the mine owners that had resisted the provision of more shafts, though it should be noted that two of the miners who died were twenty-six-year-old Robert Coulson and thirty-three-year-old John Coulson. Tragically, Clark Smith, one of the sinkers involved in the rescue efforts at Hartley, was killed in a fall of stone at Pelaw Colliery in 1910.

Wemyss Reid, chief reporter on the *Newcastle Journal* at the young age of nineteen, had only two months previously been reporting on the Prince Consort's death before news of the colliery disaster broke to the world. Later, as Sir Wemyss Reid, he recalled seeing one of the sinker rescuers lying insensible on a platform after being brought out of the mine:

Master sinker, William Coulson, suddenly realising who this sinker was, stooped down to kiss the unconscious lips of his son, and then without a word or a sign of hesitation, he calmly took his place in the loop, and ordered his attendants to lower him into the pit. None dared to stop him, for there was still the faint possibility that someone among the imprisoned miners might be still alive. But it seemed to us on the pit-heap that the brave old man [aged seventy-two years] was going to certain death, and we never expected to see him alive again when he had vanished from our sight. He did come back alive, however, and brought with him the terrible story of what he had seen. All two hundred imprisoned colliers were dead. They were found sitting in long rows adjoining the shaft. Most had their heads buried in their hands, but here and there friends sat with intertwined arms, whist fathers whose boys were working with them in the pit were in every case found with their lads clasped in their arms.

The 'miners' poet', Joseph Skipsey, described the last hours of the entombed miners in his poem 'Hartley Calamity':

'Oh father, till the shaft is rid',
Close, close beside me keeps:
My eyelids are together glued,
And I – and I – must sleep,

'Sleep, darling sleep, and I will keep';
Close by – heigh ho! – To keep;
Himself wake the father strives:
But he – he too – must sleep,

Oh, brother, till the shaft is rid,
Close, close beside me keeps;
Himself awake the brother strives;
But he – he too – must sleep,

Oh, mother dear, wert, and wert though near,
Whilst – sleep – The orphan slept;
And all night long, by the black pit heap;
The mother a dumb watch kept.

And fathers and mothers and sisters and brothers,
The lover and the new – made-bride,
A vigil kept for those who slept,
From eve to morning tide.

THE CALAMITY AT THE HARTLEY COLLIERY.—BRINGING THE DEAD BODIES TO BANK.

Bringing the dead bodies to bank. (NEIMME)

THE FATAL ACCIDENT AT NEW HARTLEY COLLIERY: ENTRANCE TO THE SHAFT, VIEWED FROM THE HORSE-HOLE.—SEE PAGE 113,   Shaft entrance. (NEIMME)

After it was realised that no further lives would be saved, master sinker Coulson and mining engineer T.R. Forster agreed on arrangements to bring the bodies to bank, having confirmed that the shaft was now clear of gas. Queen Victoria demanded that she be kept closely in touch with the rescue efforts. Reid remained a full week at the Hartley pithead with mining engineers and doctors, and witnessed the bravery of all involved trying to rescue the miners. His accounts were re-published in order to aid a fund raised for the sinkers. The news of the plight of the people from Hartley produced a flood of money from all over the country. Queen Victoria donated £200, a very unusual act for royalty to take at that time. Joseph Humble had the grim task of giving out the first payments from the relief fund to the bereaved families. Charles Dickens had visited Newcastle between 1857 and 1867. What an account he could have written, if he were minded to do so!

William P. Shield reported to the Mining Institute the following accident statistics on 31 January 1862:

> In the three counties Durham, Northumberland and Cumberland in the ten years ending 1860 there were 1,597 deaths by colliery accidents, or 159.7 per annum, i.e. a death rate of 3.13 per 1,000; 184,000,000 tons of coal was raised in ten years and 8.7 lives were lost in the production of every 1,000,000 tons of coal.

Another event was taking place in 1862 in America. Abraham Lincoln announced that by the end of that year he would issue the formal emancipation of slavery. The black slaves of America and mine slaves of Britain had carried a heavy burden, and there were further battles ahead for both before justice was achieved.

5

Slavery, Serfdom, Strikes and Sinkers

During the Egyptian and early Roman empires, shaft sinking was carried out by slaves, criminals and prisoners of war. In Roman times it was recorded that 30,000 slaves worked over ten years to construct a tunnel east of Rome to alleviate flooding. Slavery was to be the lot of miners for many centuries to follow. It is said that Caesar found the woad-covered savages of Britain 'such stupid people that they were not fit to make slaves in Rome'. During the Black Death troubles, Bishop Hatfield issued a warrant to John de Walgrave to seize workmen and coal-bearers for his needs in Durham. The pits were first in the care of the 'viewer', a title which at an earlier time may have applied to a forest official. Records show that a 'banksman', if promoted to the position of surveyor, was given the power to imprison his workmen. In the south-west of England, where mining in Britain started, the miners were given some independence, but the nature of the work carried its own burden.

The Scottish Parliament with 'deliberate intent brought serfdom into law' by means of the 1606 Act relating to 'Anent Coalyers and Salters', 'prohibiting miners from removing themselves from that occupation'. Beggars, vagrants and petty criminals were forced into lifelong bondage in the mines. Many coal owners gave 'arles' (binding money) at the baptism of children born to serfs. In Scotland and the border counties with England, the system of bondage for colliers was to last from 1606 to well into the nineteenth century. Although the colliers' wages were relatively high, their working and living conditions amounted to virtual legal serfdom; they were the property of the mine owners, whose only obligations were to provide them with basic accommodation all their life and a coffin at burial. Collier serfs were not slaves in a strictly legal sense, in that they could own and bequeath property and undertake legal actions, but this rarely happened. It was suggested that serfdom was designed to ensure a sufficient supply of labour. Diaries of earls in Wemyss, Fife, showed that serfs were also appointed as grieves. As early as 1662, there were records of miners in Newcastle considering themselves oppressed, and 2,000 miners petitioned the king complaining of their harsh treatment at the hands of the coal owners. However, the Tyneside 'hostmen' did have some goodwill towards their pitmen by paying wages 'at a rate farr above the markett price', instead of in corn.

Workers in other parts of England were showing signs of discontent, and miners in particular were starting to revolt. There was also unrest in Scotland with the Earl of Mar leading the first Jacobin uprising in 1715, and afterwards establishing coal mines at Alloa. The miners of Kingswood in Somerset were always known as a rowdy crowd, and in 1709, 200 of them marched to Bristol in protest at bread prices. They were promised a reduction in the price of wheat. At this time the Somerset weavers came out many times in protest against the introduction of looms. In 1740, when armed miners rioted in Wales at Rhuddlan over low wages, troops were called in. Even a cockfight leading to a dispute in Cardiff in 1763 caused serious rioting.

The colliers were bound to their owner for the extent of the hiring period. Under the bond system, which mainly related to the coalfields of Scotland, Northumberland, Durham and Cumberland, people were contracted to a 'Master' in return for a nominal wage – usually a shilling per week. The terms of the bond required them to work continuously at one colliery for one year, but the colliery owner gave no guarantee of continuous or, indeed, any employment at all. Anyone breaking their bond was liable to arrest and, if convicted, they could be blacklisted or 'transported'. Frequent newspaper advertisements appealed for the return of runaway miners, with the threat of prosecution for anyone who employed them. In 1703 the High Sheriff of Durham drew up a bond with the 'hweres' [hewers] of Benwell. The great strike of 1765 brought out 4,000 colliers on strike between the Tyne and Wear when colliery owners tried to change conditions of hiring. Several guineas were being offered by Newcastle pits with special hazards to attract them away from other pits. Miners including sinkers at Bushblades Colliery in 1766 were bound to George Silvertop Esq., of Stella in County Durham. It would appear that binding included sinkers as well as miners. One year earlier, the yearly bond was the cause of a long strike by Durham miners.

The Scottish Act of 1779 noted that 'many colliers and coal-bearers, and salters, are in a state of slavery or bondage', but from 1775 the need for coal was so great that all new men entering the mines were allowed to be free, and by 1799 all miners were freed from their virtual slavery. By 1808, bonds began to include an article providing for settlement of some disputes by arbitration, and a bonus of £5 was established in 1809.

At this time serfdom was abolished; records showed that 'the hewing rates of coal were 2s 11d to 3s 4d a ton in Midlothian, compared to only 1s 6d per ton in Yorkshire and 1s 1d per ton near Newcastle'. It was considered that the English collier had a greater productivity, but Scottish colliers paid their own bearers, usually their wives and daughters. At Byker Hill, Northumberland, in 1774 the price of coal was only 1s per ton, but by 1800, the amount had risen to 2–3 guineas, and in 1804 at Tyneside collieries to 12–14 guineas and 18 guineas on the Wear.

Arthur Young declared in 1771 that 'every one but an idiot knows that the lower classes must be kept poor, or they will never be industrious'. Early Methodism only reinforced this attitude to the education of working people, but this did not last since later many became union leaders despite their Methodist upbringing. Most eighteenth-century workers would have gladly exchanged their industrial employment for a month of harvesting, especially during the warm weather, and they remained suspicious of advances in technology which required them to give up their traditional ways of working. In 1776, a cast-iron tramway nailed to wooden sleepers was laid down at the Duke of Norfolk's colliery near Sheffield. The designer of this early example of railway, John Curr, had to hide in a neighbouring wood when the colliery labourers rioted and tore up the railway and burnt the coal-staith. This Luddite-type response was widespread, but thirteen years later engineer William Jessop was recorded as constructing a railway in nearby Leicestershire without interruption. Gradually these railways were found in all mining districts, and alerted mining engineers to the powers of the steam-engine as a motive power in their industry.

Samuel Smiles noted that:

The pitmen, or 'the lads belaw' are a people of peculiar habits, manners, and character, as much so as fishermen and sailors, to whom, indeed, they bear in some respects, a considerable resemblance. Some fifty years since, they were much rougher and worse educated class than they are now; hard workers, but very wild and uncouth; much given to 'sleeks', or strikes.

A sort of traditional disrepute seems to have clung to the pitmen, arising from the nature of their employment, and from the fact that the colliers were the last class enfranchised in England, as they were in Scotland, where they continued to be bondmen (bound to the colliery owner) down to the end of the eighteenth century.

The coal mining communities were generally regarded with some distrust and suspicion by other workers and were excluded from many areas of social contact. At the first Durham Miners' Gala, held in 1871, the shopkeepers of Durham boarded up their properties, such was the dread of pitmen, especially when they gathered in large numbers, seeing as they had done so through the years to protest at their treatment by the colliery owners. The coal miners had always been at the forefront of protesting at inhuman working conditions. Coal created a new gulf between classes. The medieval peasants and artisans, whatever their disabilities and trials may have been, were not segregated from their neighbours to anything like the same extent as were coal miners of the seventeenth century in most colliery districts. Children had been forced to work in mines from as young as three years old, and this was accepted by their parents as a necessary contribution to the family's survival. It was also considered by both owners and miners that children introduced to the coal mine at an early age would find it easier to get used to their work underground. This also applied to boys employed in sinking operations. In 1881 Ralph Hann, aged only fifteen, worked as a sinker at Whitburn, and at Llanwonnon, Glamorgan, in 1871, sinker Rees Davies was fourteen years old.

The laws to protect property had always been severe, with punishments ranging from transportation to death. In 1736 it was made a felony to set fire to a pit. Small-scale industrial strikes took place in Gateshead in 1740, 1765 and 1784. Even seven years after the first strike, discontent was still rife, with owners advertising for the arrest of incendiaries who were burning the pithead machinery. Riot and destruction swept along the whole of the Tyne valley in 1765, and this affected not only the local miners, sailors, keelmen and staith men, but also spread to the London coal-whippers. Despite the Malicious Injuries Act of 1769 to prevent damage to pit engines, surface winding gear was being destroyed at colliery after colliery. The coal owners were afraid that competition for pitmen would drive up wages, or the men would be enticed from the deeper and more dangerous pits.

Miners were accused of being followers of the revolutionary doctrines of Thomas Paine. To avoid prosecution under the Combination Laws, miners formed friendly societies rather than unions. The Government feared general insurrection as riots spread across the country. In view of the ruling classes' distrust of local militia, the Home Office in 1793 urged that in Northumberland and Durham they be replaced by regiments from southern counties. Starvation was decimating workers in south-east England, and food riots in France and across Europe continued up to 1800. Miners in counties with no previous history of protest were beginning to fight back. In 1783 lead miners in Aberystwyth rioted over high corn prices. Nine years later some 4,000 colliers marched to protest about their wages. Unusually, their demands were met.

During the first great strike in 1810 led by the miners' union, its members were solemnly pledged to secrecy under penalty of being 'stabbed through the heart' or 'having bowels ripped out'. This strike lasted for seven weeks. Durham Jail became so full of dissident miners that the Bishop offered the use of his stables to house others who were arrested. In the absence of magistrates the Church played a leading role in deploying police constables. The misery suffered by mining families was immense, but the strike in the end broke down due to famine and evictions. The miners eventually returned to work in 1811. This was the time of the Luddite attacks which started in Nottingham. The end of the war with France in 1815 led to a dramatic drop in the demand for manufactured products, and this resulted in many ironstone miners and colliers being thrown out of work.

The unrest in the North East also spread to the transport of the coal to the staithes, and this seriously affected the export of coal to London and other ports by means of the sea-going colliers. The coal owners began to consider other options for the export of the coal. Also at this time, the accelerating development of the railways was due to the increasing congestion of Tyne, Wear and Tees, and the keelmen feared for their future. At Sunderland Drops, keelmen demolished the staithes, setting them on fire. One man died by falling timber, and dragoons from Newcastle were called to disperse the crowd. However, the keelmen, who were the kingpins of the coal trade, did not appear to be attracted to trade unionism, opting to retain their independence in their battle for survival. As well as in 1819, the 'keel lads o' coaly Tyne' went out on strike in 1844 on the issue of 'overmeasure' where the keelmen were paid the same amount even when the hostmen to avoid tax had increased the size of the ship hold by 26.5 tons.

Fearful of a revolution, the Government, using the Combination Laws, reacted swiftly and harshly to workers' 'illegal' gatherings, sometimes known as 'brotherings'. The Tolpuddle Martyrs were made an example by being transported to Australia in 1834. William Cobbett, the champion of farm workers, during a visit to Sunderland in 1832, considered that:

> … the working people [here] live well … The Pitmen have twenty-four shillings a week; they live rent-free, their fuel costs them nothing, and their doctor costs them nothing. Their work is terrible, to be sure …but … their houses are good and their furniture good.

His views reflected the deteriorating conditions of agricultural workers at this time. There was a continuing drain of workers from the countryside into the towns throughout the nineteenth century, attracted by better pay: 'Collier lads get gowd and silver, Factory lads get nowt but brass …'

Generally, miners' leaders tried to reach some accommodation with the mine owners by setting out their grievances to the viewers. In a communication dated 14 January 1826 from the Cock Inn, Head of Side, Newcastle, the miners drew attention to the injustice of the bond. They received no response to this and further requests for conciliation. Severe riots took place in 1831 when 1,500 miners stopped the pits working in the Bedlington district, throwing the corves down the shaft at the Glebe Pit. The following month mine workers refused to sign the annual bond which had expired on 5 April. At Hetton Colliery, miners' leader Thomas Hepburn founded the Northern Union of Pitmen in 1831 – 'The Pitmen's Union of the Tyne and Wear'. He wanted a reduction of working hours for boys from 16 to 12 hours a day and an abolition of the 'Tommy Shop' system, and led his members on strike. At this time the owner of the Bedlington ironworks was concerned about the effect of strikes on his works since at the nearby colliery a mob had sacked the viewer's house at Cowpen wrecking the winding gear. He contacted the Prime Minister, Lord Melbourne, and informed him that 1,000 pitmen had attacked Netherton Colliery, destroying the steam engine, and forced all pitmen to join the strike. The Mayor of Newcastle was also in touch with Melbourne about the recent strikers stopping work at Jesmond Colliery, with corves and ropes thrown down the shaft. In many disputes the shaft often became the focus for acts of violence. By April 1831 there was not a single colliery in operation between the Tees and the Tweed.

The first signs of cholera were noticed at Sunderland in late 1831. Not far away at Callerton, Coxlodge and Waldridge, lead miners were being imported, and the local pitmen refused to work with them. Miners were able to earn more in North East collieries than in lead mines. At Waldridge the angry coal miners stopped the pithead engines, endangering the lives of the lead miners working below, and several rioting pitmen were put on trial. This became known as the 'Waldridge Outrage' of December 1831 when striking miners dumped various gear down the mine shaft, stopped the pumping engines, and threatened the lives of several non-union workers at work down the pits. The Government offered a reward of 250 guineas and a free pardon to accomplices giving evidence, and a further 250 guineas reward from the colliery owners. Seven men were put on trial and six imprisoned. This was the culmination of a series of disputes dating back to the opening of this colliery in 1830.

In Wales, the 'Merthyr Rising' was the culmination of miners' frustrations with their employers over working conditions. At this time in the industrial valleys of South Wales, when the ironmasters brought in the blacklegs, they were treated with equal savagery by clandestine organisations known as 'Scotch Cattle'. In 1831 Richard Lewis was hanged in Cardiff and three years later Edward Morgan was hanged in Monmouth Prison. Using bull's head and horns as a symbol, the miners hunted down the 'traitors, turn-coats and others'. In Staffordshire at this time, the potters supported the colliers' strike 'in defence of their just rights'. One of their main grievances was the owners' continuing use of their tommy shops which continued to affect the iron and coal industries. Even though the 'Truck Acts' of 1820 and 1831 had outlawed their use, the practice remained in place in some areas into the 1870s.

In February 1832, William Coulson, aged forty-one, was making his name as a master sinker, sinking shafts in the Chester-le-Street area. At the same time in Ipswich another William Coulson, aged seventeen, was sentenced to seven years' transportation to Van Diemen's Land (now Tasmania). His crime was stealing a firearm from a dwelling house; his trade was 'bread and biscuit baker'. As he waited on the hulk off Portsmouth, he was given six days on the treadmill, and received twenty-five lashes. He sailed in the *Lord William Bentick* with 186 male convicts. After twelve years' hard labour in Australia he married Ellen Mary Reilly from Dublin who also had been transported for seven years for stealing a cloak and shawl from her mistress.

In these revolutionary times, the Government was more wary of confronting the miners, hoping that they could leave it to the colliery owners. In May 1832 there was a serious riot at South Shields Colliery due to an attempt to prevent the bound men from going to their works. Forty-two lead miners from Cumberland brought in by the owners were pelted by the local miners, with two men seriously injured. When miners refused to work and leave their cottages, special constables had to be sworn in to deal with the emergency. Several families were evicted from their homes. The Rector of Gateshead appealed to the Mayor of Newcastle, resulting in reinforcements firing on striking miners. The mine owners' aim was to smash unionism in the North East. The next morning a number of pitmen congregated at Hebburn Colliery and threw down the shaft all the corves, rolleys and loose material to the great terror of the men below. The pitmen had generally resumed their work by mid-June, but peace did not last long when a great riot took place at Friars Goose. A few days later magistrate Nicholas Fairles Esq. of South Shields on his way to visit the viewer regarding the on-going strike, died after being attacked by pitmen William Jobling and Ralph Armstrong. These men were hanged and left to putrefy as a warning to the miners.

A few colliery owners gained the respect of their miners, such as Lord Durham, who was nicknamed 'Radical Jack' for his liberal views. During the long strike of 1832 in Northumbria, the colliery owners recruited strike-breakers from the metal mines, and from other coalfields, Staffordshire again supplying more than other districts. However, forty-eight Welsh miners who had just arrived at Ouston left their employment and joined the strikers. Retaliation came in the form of evictions and detachments of the new London police, and the Queen's Bays, were posted around the town to quell any disturbance. The miners themselves were armed, and shots were exchanged on both sides before evictions could take place. After this strike, five of the seven men transported to Botany Bay from

Hurriers at a Yorkshire colliery in 1842. (Parliamentary Archives)

Jarrow were Primitive Methodists. Also, nine of the eleven jailed from West Cramlington were local preachers. It was reported that the fracture of a shaft rope at Thornley killed two blacklegs. At Eppleton Colliery, the company was having to cope with a massive water feeder of 5,000 gallons per minute. When the sinkers joined the strike against the bond in 1832, the mining engineer and viewer John Buddle was faced with the problem of there not being enough men to fight the water. This is one of the few records of sinkers becoming directly involved in industrial action.

The Female Political Union of Newcastle in 1839 argued that miners' wages were so low it was forcing their wives and children into degrading labour. The Earl of Shaftesbury (an appropriate title for someone who was to be concerned by the deaths of children down colliery shafts) was successful in passing the 1842 Mines Act which forbade the underground employment of women and of children under ten years of age. These measures would have had the greatest impact on trapper boys, whose ages ranged from four to nine years old. In most coalfields an attempt was made to minimise the use of labour by introducing tramways and winches; in Scotland, coal was moved underground on the backs of women and girls. Some young girls 'hurried' or pulled tubs of coal and they were effectively naked when their breeches became torn through wear and tear. When asked by a commissioner where he would go to when he died, a young boy replied 'buggery'. He had heard his father say that miners had been 'blown to buggery' in an underground explosion!

Sinking in 1840 at Thornley Colliery reached a seam of first-class quality, and the next year, boring was used to prove its quality. However, this good news was quickly forgotten when Thornley became the first colliery in east Durham to be affected by the industrial unrest in 1843. At this colliery the fracture of a shaft rope resulted in the deaths of two blacklegs and caused 'great shouting and marks of rejoicing' from a 'mob, consisting principally of woemen'. In contrast, when Staffordshire miners arriving at Marley Hill found they had been duped by the recruiting overman, John Greenhorn, they returned home and vowed that if Greenhorn approached them again they would 'mark' him. Around the same time, the Government was trying to cut the tariff on sugar imports which would have encouraged plantation owners to go back to employing slaves.

The Great Strike of 1844, led by Thomas Hepburn, was driven by a main grievance of poor mine ventilation and the lack of efficient safety equipment. The weight of the British press was against the miners, and owners arranged for coal to be imported from Scottish ports into the North East, which might be the origin of the expression 'to carry coals to Newcastle'. The new east Durham pits tended to be more radical and unionised than the long-established pits, with Hetton the epicentre of trade unionism during the first half of the century. There had been a long tradition in the iron mines of Cumberland, Egremont to Cleator Moor, of importing miners from Ireland, County Durham and Isle of Man, but in the North East, newcomers were not generally accepted by the indigenous miners. The local miners at Seaton Delaval in Northumberland convinced the blackleg Welsh miners that they were not welcome, and persuaded them to return home. A poor Welshman called 'Blind Davy', who was near-sighted, was met with 'seven or eight picks on his back, and with these he went right to the pit mouth, fell to the bottom, and was killed'. After the strike the tools of the blacklegs were thrown down the shaft, and wires were stretched across underground roadways to catch the heads, throats and bodies of Welsh blacklegs. Those remaining were thrashed by the native miners and sent back to the 'land of song' with the following reminder in their ears!

The Blackleg Miners

Oh, early in the evenin', just after dark,
The blackleg miners creep to wark,
Wi' their moleskin trousers an' dorty short,
There go the blackleg miner!

They take their picks an' doon they go
To dig the coal that lies belaw,
An' there's not a woman in this toon-aw'
Will look at a blackleg miner.

Oh, Delaval is a terrible place.
They rub wet clay in a blackleg's face,
An' roond the pit-heaps they run a foot race
Wi' the dorty blackleg miners.

Now, don't go near the Seghill mine.
Across the way they stretch the line,
Te catch the throat an' break the spine
O' the dorty blackleg miners.

They'll take your tools an' duds as well,
An' hoy them doon the pit o' hell.
It's doon ye go, an' fare ye well,
Ye dorty blackleg miners!

Se join the union while ye may.
Don't wait till your dyin' day,
For that may not be far away,
Ye dorty blackleg miners!

During the early 1850s massive deposits of high-quality coal were discovered in the Rhondda valleys and special sinkers were brought in from Llansamlet. At this time apprenticed paupers were being imported from Temple Cloud in Somerset. Colliery owners in North Wales cleverly used earlier xenophobic tensions to raise suspicions that miners from Staffordshire were taking over local jobs, which had led to local miners being turned out of their homes. Also at this time, after a long and acrimonious strike of thirteen weeks at Methley Junction Colliery, Yorkshire, the following letter from a miner was received by the coal mine proprietor Henry Briggs. This gives a good idea of the fractious relations between this owner and his miners:

Mr Briggs
I will tell you what i think by you About this struggle. You are getting
an ould man and besides you are a tyrant – ould B ---r now Sirs what
do you think to that bit – we have stopped 13 weeks All redy, but i
have myself sw to take your life and your son also. But you shalt not
live 13 days. Depend on it my nife is sharp. But my Bullits is shorer
than the nife and if a i can under the time i will by God – if it be at noon
day wen i see you shall have the arra if it be in your Charrit. Like ould
Abe Now reade that and pray to God to forgive your sins to be reddy.

There was a revival of trade unionism in the 1860s, but like Thomas Hepburn, Martin Jude, who had fought tirelessly for miners' rights, died in poverty. He did not live to see the 1861 Mines Act which governed safety and inspection of mines, and the transfer of the position of check weighman from the owners to the miners. However, the Durham pitmen had to accept the reintroduction of binding, causing the Northumberland miners to secede and form their own association. In 1866, following a year of strikes for higher wages, the owners brought in 400 Cornish miners and their families, and these strikes petered out. Some of the small owners let the men return at much the same rates as before the dispute, whereas the large companies like Wigan Coal and Iron Company which had invested large amounts of capital were able to hold out for a full 15 per cent reduction. Similarly during the 1884 strike the small owners were aware that if the pits closed they were unlikely to reopen. The position of these owners was strengthened when they acted in combination.

In 1863 a 'riot' took place near Chesterfield in Derbyshire involving violence between English and Irish workers employed at Sheepbridge Colliery. As well as racial tensions, disputes often arose due to the suspicion that incoming workers were receiving preferential treatment from the colliery owners. At Brancepeth in Durham, the 'rocking strike' resulted from the need to rock the filled tub to avoid deductions made to miners' wages. In 1866 a disagreement took place in the Miner's Arms at Baybridge near Blanchland, Durham, between the 'cornies' and the locals. During this decade, when Wheatley Hill pit was being sunk, the local farmers did not take kindly to the sudden invasion by sinkers into their area due to start cutting the turf and begin their operations. The sinkers were driven off until the bailiff intervened.

Little mercy was shown to infringements of the law by working people. Middle-aged Irish woman, Margaret Ryan, living near St Helens in Lancashire in 1870, was charged with stealing 24lb of coal from Groves Colliery. Colliery overlooker Joseph Heyes reported that at 8.15 a.m. he saw her on the coal wagon taking coal. She ran away but he caught her. She told him that it was a very small piece of coal, and explained to the bench that her husband had been out of work since Christmas, and she had taken the coal to make her children warm. Even though there was nothing against the woman's character, she was sent to gaol for seven days. At that time it was unlikely that her husband received any relief, especially if he had not been born in the parish.

In 1872, the union, Durham Miners Association, achieved the removal of the yearly bond, but some strike leaders remained blacklisted and had to move away to obtain employment. So poverty had induced the blacklegs to move in, and poverty had resulted in the blacklisted being forced out. The unions' struggle for legal status and free bargaining continued, but miners mostly were not attracted by the dominant role of the physical (New Model) way forward. A Durham coal miner, Thomas Burt, entered Parliament as Radical Labour MP for Morpeth in 1874. Therefore it was not surprising that Keir Hardie, a former miner, was to become the first leader of the Labour Party in 1900. A religious man not of Catholic persuasion, Hardie nevertheless was a strong advocate of Irish Home Rule. Perhaps people were beginning to believe that a new kind of life was possible. However, one year earlier, following the importing of Irish and other blackleg strike-breakers by the Scottish coalowners and ironmasters, the strike in Hamilton collapsed and the twenty-five-year-old Hardie was expelled from Lanarkshire, forcing him to find pit work in Ayrshire.

When the new coalfield of Kent opened up, there was a heavy demand for labour, and miners' past working history was not scrutinised. In 1921 Pearson, Dorman & Long (PDL), the owners of Betteshanger, had combined with the intention of creating a great coal, iron and steel industry in Kent. After the 1926 long lockout of miners and the resulting high unemployment, particularly militant miners were drawn to Kent where they found employment. Before the miners arrived, roving teams of sinkers sank the pits. Most of the sinkers were from the north of England and South Wales. Billy Marshall, a miner at Betteshanger, recorded his experience:

> At the time I started at the pit I suppose there were as many sinkers as there were colliers. Now the sinker was a drifter, he was a rough hand, and I remember the first Saturday night I was in Deal I saw two of them stripped off to the waist, fighting bare fist in the churchyard in the High Street. Now just imagine the impression it would make on a seaside town in Kent, that kind of episode, and so that all miners were of the same category – the sinker type.

In the twentieth century, miners were to continue to play a prominent part in British trade unions and politics. In combination with other unions, miners played an important role in the foundation of the first Labour Government in 1924, and during the general strike two years later nearly brought down the Conservative Government.

In other European countries, miners were at the forefront of political movements. The Asturias miners' strike in 1934 led to the Spanish miners in open warfare against the government of General Franco. 3,000 miners were killed in action, with a similar number taken prisoner and thousands losing their jobs. The final nail in the coffin for the British coal industry was the miners' strike in 1984, when many miners believed that 'Thatcher sank the pits'! As during previous conflicts, miners' wives and families gave unwavering support to their husbands and fathers.

Religion, Chartism and the Irish

Particularly in the nineteenth century, religion played a large part in the mining community, and it is not surprising that when an engineman died suddenly at work, the inquest should conclude that he had 'died by visitation of God'. Like most of the mining communities, the industrialists forcing the changes in the North East were mainly of dissenter inclination, for example the Quaker family Pease. The established Church kept at arm's distance the poorer section of the community and therefore the mining families were drawn to the Methodist preachers who considered the poor deserved as much attention as the rich.

In the late 1700s in Cornwall, as in South Wales and the North East, there was conversion in droves to Methodism, so that by the 1850s it was the principal denomination. Initially, it was the 'poor and vulgar' that were attracted to its teachings, and gradually the more prosperous citizens such as merchants and mining agents began to attend chapels rather than churches. For a few miners, not under the influence of the 'Society for the Suppression of Drunkenness', after their shift they would visit the many 'winks' with bouts of hard drinking known as 'choruses', usually on a Saturday evening. It was known for hymn singing to take place down the mines.

In his journal of 1747, John Wesley commented that after preaching at South Biddick he rode to a village called 'Renton' (Rainton in County Durham) where there were many collieries and an abundance of people. Four years earlier he had preached to 10,000 tin miners at Gwennap Pit in Cornwall. In contrast to Cornish miners, Durham miners were not normally superstitious and generally 'solid' in their outlook. Their teaching of the need for strict discipline in their private and working lives would have appealed to both miners and employers. The Methodist or Ranters Revival in colliery districts did much to improve the coalminer's image. For example in Cornwall, barbarous fighting contests had greatly reduced. Samuel Smiles mentioned that the head viewer, when looking for George Stephenson, knew that he would find him at 'the preaching' in the Methodist chapel. Later, Methodists gained the reputation of holding radical views.

The massacre of Peterloo in Manchester culminated in the greatest rally recorded at Newcastle in October 1819 on the Parade Ground, Percy Street. The principles of Radical Reform were spread and espoused 'with all the fervour of a moral or religious feeling'. In 1830 'gins' removed from Whitehaven pits during 'Swing' rioting, gave a hint of the Chartist agitation about to commence. The Chartist events held in South Wales in 1839 were to claim a place in history. Starting in 1821, Monmouthshire colliers went on strike, and in 1831 Merthyr Tydfil rioters were assisted by local ironworkers and colliers. A local resident, John Frost, formed a Political Union. Eight years later both John Frost and Zephaniah Williams were leaders in spreading Chartism in Blaenau Gwent. When matters came to a head in 1839, over 40,000 Chartists, including colliers, attended what has been considered the largest gathering in the nineteenth century. Frost and Williams were found guilty of high treason and condemned to death, although this was commuted to life transportation to Tasmania. Frost was to return to Britain in 1856 but Williams was to die in Australia in 1874.

With lack of trade union success, miners were turning to political alternatives. Serious disturbances took place during the Chartist Riots in 1842 when 10,000 marchers descended on Wigan and stopped 3,000 locals from working in the mills and collieries. In Staffordshire, William Cooper, a Methodist preacher, was sent to prison for reading the following poem and supposedly inciting the potters and colliers on strike at Hanley to riot:

SLAVES, toil no more! Why delve, and moil, and pine,
To glut the tyrant-forgers of your chain?
Slaves toil no more! Up from the midnight mine,
Summon your swarthy thousands to the plain;
Beneath the bright sun marshalled, swell the strain
Of liberty; and, while the lordlings swell view

Your banded hosts, with stricken heart and brain,
Shout, as one man,–'Toil, we no more renew,
Until the Many cease their slavery to the Few!'

The aftermath of the Great Famine in Ireland in the mid-1800s was to create a flood of Irish families to all areas of mainland Britain, including the North East. The influx of the Irish was regarded with some alarm by the local population on account of employment and the fears of 'Popery'. Blacklegs were considered with dread, as these incomers could take locals' jobs and their livelihood. In particular, the Irish seemed to be spurned since they also represented immigrants that held fiercely to a different religion, and their large families gave rise to the Malthusian fear of overpopulation with the need for more parish relief. When Lord Londonderry brought in a hundred blacklegs from his estates in Ireland to his collieries near Rainton, his overviewer concluded that the 'Irish had not been the mindless serfs he had hoped'. Thomas Carlyle, normally not generous with his views, noted 'The postman tells me that several of the poor Irish do regularly apply to him for money drafts and send their earnings home. The English, who eat twice as much beef, consume the residue in whiskey and do not trouble the postman'. In 1836, a Catholic priest admitted that the Irish workers were 'more prone to take part in trade unions, combinations and secret societies than the English'. This was not the case when starving Irish first landed on the British mainland. Ironically, the coal owners, by introducing strike-breakers from Ireland, were perhaps sowing the seeds of greater resistance from their workers in the future.

Commissioners in 1842 at Walker Ironworks commented that the Irish were not skilful or ambitious, but witty and good-tempered. It was well known that Irish labour was used in the 1840s to build the railways. There followed a period of unemployment during which a number of Irish turned to highway robbery, some forming bands of footpads (common thieves preying on pedestrians). Irish robbers seem to have been the terror of the citizens of Durham City who had to walk the roads to the nearby villages of Shincliffe, Gilesgate and Neville's Cross. With the sinking of the shafts in mid-Durham, for example in Esh, the Irish were drawn to more legitimate ways of earning. Working-class prejudices towards the Irish were beginning to lessen. A report on the sinking of Maltby Colliery in Yorkshire in 1908 noted that the sinkers were a special breed, often working seven days a week until the job was done. They would never lose a shift for perhaps a month, and then 'they would be on the whiskey'. They were known as men of great skill and daring, seeming to have no fear of shaft work. Most of the sinkers who carried out the heavy labouring involved in sinking were Irish.

The miners' union had an uphill task in motivating their members to continue the struggle for decent working and living conditions. In 1864, the National Union of Miners delegate E. Rhymer, from Spennymoor, Durham, spoke gloomily of the 'ignorance, cowardice, and drunken habits' which made some pitmen accept without resentment wages which fell as low as 2s a day, and which never exceeded 4s 6d. Poverty, stupidity and greed made them dependent on the wages of children who worked 12, 14 and 16 hours a day. When the new Pitmen's Union, the Durham Miners Association, was formed in 1869, it was led by men showing missionary zeal, with the Primitive Methodists putting God on the side of the miners. By the 1880s the Irish were an integral part of the Association, and 10 per cent stood at the lodge elections in 1885. It would seem that the fate of Methodism and mining has been similar in mining areas such as Cornwall where the landscape is scattered with derelict chapels and engine houses. However, like Methodism, the legacy of the coal miners is not only their material remains but the inspiration they gave to fight for a more just society.

6

Miners, Master Sinkers and Mining Engineers

Mine workers have always been a close community since their lives and livelihood depended on it. Over the ages miners and their families have been treated by the general public with suspicion and sometimes fear, similar to the response that gypsies and travellers have often received. At the commencement of colliery works, pit sinkers were the first outsiders to suddenly invade an area, and like the navvies they gained a reputation as strong workers, but were also known for heavy drinking and rowdiness.

Well-sinking to obtain water would have been a familiar activity in many villages, but outside the world of mining, the work of the pit sinker was generally unknown. Sinkers sank the shafts to provide access to the coal seams, thus allowing coal extraction from the colliery to commence. At early times a banksman at the surface would have operated a windlass or gin to wind up or down the shaft, the sinkers and their materials, attached to a rope or in a wicker basket. Later the sinker mainly carried out his work in the shaft from a kibble (metal bucket) or working platform suspended from the winding rope or chain, and therefore his life depended on the brakesman and engineman for the safe operation of the winding engine located at the top of the shaft. Samuel Smiles, in his book *The Life of George Stephenson and his son Robert Stephenson*, noted that 'among the upper-ground workmen employed at the coal-pits, the principal are the fireman, engine-men, and brakesmen, who fire and work the engines, and superintend the machinery by means of which the collieries are worked'.

The work of the master sinker who was in charge of the sinking operation required him to be competent in many aspects of engineering, therefore it is important to examine the evolution of the 'engineer' in mining. The occupation of the engineer has been closely linked with mining from earliest times, such as military engineers who were involved in building temporary bridges and besieging castles. The miners under their control used their mining and engineering skills to drive tunnels under the fortress walls. The word 'engineer' (from the French *l'ingénieur*), sometimes misspelt in census records as 'ingineer', was derived from the word 'genius' – the innovation and ability to design and resolve problems. This started with the invention of basic tools such as the wheel and the lever. Mining has strived to develop machines to increase the power produced by humans and animals. The engine came to be the name used to describe any such contraption, sometimes worked by a horse, and thus horsepower became a unit for measuring mechanical effort. An early winding mechanism for moving men and materials up and down a shaft using a horse was known as a horse [en]gin. The 1851 census recorded the occupation of Hunerey Handra, aged fifty, at Gwinear Cornwall as an 'Ingin Man', and in 1891 Charles Salter, aged thirteen, was described as a 'Jinny brake boy at colliery' in Barnsley, Yorkshire.

Mining engineers and master sinkers at Whitworth Colliery, Durham, in 1860. (Beamish Museum)

Man's ingenuity in designing the steam engine was a critical factor in advancing the Industrial Revolution, and the mining industry was to make a major contribution to spreading its application to many fields of engineering. Coalfields throughout Britain started to use steam-powered pumps from the early part of the nineteenth century, and the Northumberland and Durham coalfield was one of the first to recognise the advantage this would give them. This region has always had a thriving interest in applied science as shown by the formation in 1793 of the Newcastle Literary and Philosophical Society, well patronised by mining officials. George Stephenson demonstrated his modified safety lamp to its members in 1815, and in 1880 the society's lecture theatre was the first public building to be lit by electric light during a lecture by Joseph Swan. The society attracted to its early meetings the well-respected mining engineers, John Buddle and Nicholas Woods, as well as civil engineer William Fairbairn, friend of Robert Stephenson, who began his career as an engine apprentice at Percy Main Colliery near Newcastle. However, colliery owners initially showed little interest in providing even basic education for their manual workers. Miners in endeavouring to improve their working conditions, particularly during the nineteenth century, were instrumental in making basic education available to future generations of working people.

The Industrial Revolution gave birth to professional engineers – mining, mechanical and civil – and then it was possible for engineers to span these engineering disciplines. John Smeaton from Leeds described himself as a civil engineer in 1768, as distinct from a military engineer, and thus coined the name of this new profession. Nine years later he was designing a water gin (basic winch powered by water) for a coal mine near Newcastle. He stated with pride that 'Civil engineers are a self-created set of men whose profession owes its origin, not to power or influence, but to the best of all protection, the encouragement of a great and powerful nation'. In theory these aspirations could have applied to all these new emerging professional engineers, but in practice commercial interests always played a part. 'Canal Mania', the rapid construction of canals from 1790 to 1810, was the principal reason for the emergence of civil engineers as there was a great need for specialists to manage this massive venture. Similarly, with the start of deep mining and the development of large mining operations in the North East, the specialist skills were provided by properly trained mining engineers. James Brindley, a civil engineer born in Derbyshire, who gained a reputation for major canal works, apparently complained of having to combine the roles of 'land surveyor, carpenter, mason, brickmaker, boatbuilder, paymaster and engineer'. These canal works as well as the later railways required the sinking of shafts in the access tunnels.

During the nineteenth century, engineers were more multi-disciplined than today. In the census records many of these engineers described themselves as 'mining & civil engineers'. It is clear that the Stephensons and Brunel were as much mechanical engineers as civil engineers. The mechanical engineers as a professional body broke away from the civil engineers and formed their own institution in 1847, and it is not surprising that their first two presidents were railway engineers George and Robert Stephenson. Towards the end of the century a new engineering professional, the electrical engineer, was being sought after by collieries as electrical power was taking over from steam. Thomas Tredgold, born at Brandon, County Durham, in 1788 (of Cornish extraction) was a pioneer in steam and railway engineering. He will be remembered by civil engineers for his quote which has defined civil engineering ever since: 'Engineering is the art of directing the great sources of power in nature for the convenience of man.' Like the author of the 'Compleat Collier' a century before, Tredgold used the word 'art' to describe the know-how and skill based on good practice required in carrying out engineering works, and the need for proper planning and organising of the work. Simply, a good engineer needs to be 'rounded' in all aspects of the works.

In Northumberland and Durham, the mines were in the ownership of large proprietors who had exclusive rights to minerals which lay beneath their lands. They came from mainly aristocratic and middle-class backgrounds and had an enormous influence in their own coalfields and beyond. In 1829, only five out of the forty-one collieries on the Tyne were worked by the proprietor, the rest being in the hands of lessees or adventurers. Although there was great rivalry between the owners, they were not slow in combining their industrial muscle to resist miners' attempts by means of strikes to improve their working conditions. These entrepreneurs were prepared to invest large sums of money in the hope of finding large seams of 'black diamonds'.

Traditionally, colliery shafts or pits were named by colliery owners using their relations' first names; occasionally, pits were named after master sinkers who sank them. For example, Thomas Fox, known as 'our old sinker' at Hainsworth in Cumberland, was noted as having a pit named after him: Fox Pit at Greenbank, Whitehaven. Strangely, a master sinker Thomas Harrison in 1867 gave a

William Shields, master sinker. (Beamish Museum)

Colliery manager, staff and workers at Lintz Colliery, Burnopfield, Durham, in 1881. (Beamish Museum)

Inspection group including Mr Villiers, Lady Margaret Arnherst, George Robinson (certified manager), William Stephenson (master wasteman), Lady Alicia Arnherst, D. Selby Bigge and George May at Harton Colliery, Durham. (Beamish Museum)

Sinkers and colliery engineer at Browney
Colliery near Langley Moor, Durham, sunk
in 1871. (Beamish Museum)

shaft the name of 'Bang Up Shaft' after the stagecoach which he used between London and Birmingham. Scott's Pit in South Wales was named after entrepreneur and colliery owner John Scott, and in Gloucestershire many shafts took the name of the chartermaster. Mine owners took a keen interest in the practicalities of the sinking process, the resources required and of course the potential profits to be made. They were always on the lookout for the best master sinkers.

In the northern coalfields a hierarchy developed on both sides of the industry for management and workers with regard to the deployment of human resources. Working directly for the owner were the colliery officials such as managers, mining engineers and surveyors, viewers, and clerks, who were collectively called agents. The underground 'manual' workers included sinkers, colliery engineers, overmen, deputies and miners of various descriptions. Master sinkers were classed as employers when leading their own contracting company, but were not above carrying out manual work themselves. Hewers sometimes engaged and employed the putters who dragged the tubs or rolleys. Above ground, the firemen, enginemen and brakesmen who fired and worked the engines, and controlled the headstock machinery, were critical to the working of the collieries. Lines of responsibility were rigorously enforced. Later, this working structure was to be adopted outside mining by the construction industry, with the client employing resident engineers, surveyors and clerks of work, and the contractor engaging site agents, engineers and quantity surveyors to manage the execution of the works. Experienced men were always used in the sinking of shafts and cutting of roadways underground. Sinking contractors were generally employed directly by the colliery owners or managers and worked independently of the other colliery workers.

From the late eighteenth century the pit owners employed a manager with a body of trained viewers. The 'viewer' was the manager's 'eyes and ears' who took care of his mining investment by daily inspections, with particular responsibility for 'free communication of air through all the works'. Colliery viewers were trained as mining engineers, with under-viewers and apprentice viewers obtaining practical experience before promotion to the senior position. They often had a background in surveying in the coal industry, and were sometimes introduced to this position at an early age. For example, the 1841 census recorded John H. Ramsay as a colliery viewer at the age of fifteen living with the respected Durham viewer George Hunter (aged forty-five). Ten years later Parkin Jeffcock, another young colliery viewer articled to Hunter, was intending to go to Oxford University but after training at the College of Civil Engineers at Putney decided his future was in mining. In 1857 he became a partner with J. T. Woodhouse, a mining engineer and agent based in Derby. Four years later Jeffcock's bravery was recognised when he attempted to rescue men and boys trapped in a coal pit at Clay Cross during an inundation. He also worked overseas, and was involved in examining and reporting on the Moselle coalfield near Saarbrück in Germany. In 1866 he learned that the Oaks Pit near Barnsley was on fire, and during his rescue efforts he was killed following an underground explosion.

Many viewers ensured that their sons entered an apprenticeship in mining engineering as the first step towards rising to the position of viewer, and later apprenticeships in mechanical and electrical engineering became available. In line with the advancement of industry in a country previously dominated by agriculture, landowners and farmers also saw engineering as respectable profession for their male offspring. Therefore in a coal-mining area, a managerial position in the new wealth creating coal mines was an obvious choice. William Hopton's book *Conversation on Mines*, first published in 1864, included an example of an examination question presented to a young mining engineer, which tested them on their knowledge of sinking. Thus knowledge of sinking was considered an important part of the mining engineer's training.

The Hewitts are an example of the progression in mining of a father and son born in Derbyshire; both started work in the pits as miners. The son became well known in the Midlands as the originator of practical devices in connection with pumping machinery, and of improved methods of working coal on the Longwall system.

	Thomas Hewitt (b. 1814)	John Richardson Hewitt (b. 1842)
1841 census	miner	
1851 census	colliery agent	scholar
1861 census	mining engineer	coal miner
1866		assistant mining engineer
1871 census	colliery manager	mineral and land surveyor
1881 census	mining engineer	civil and mining engineer
1891 census	mining engineer	civil and mining engineer

The life of John Hewitt gives a picture of the multi-discipline engineer in the nineteenth century. Like Brunel he died at the early age of fifty-one, perhaps also due to over-working. In 1878 he set up a practice on his own account at Derby as a civil and mining engineer. Among the several important works he was in charge of during the next fifteen years were the construction of a branch railway on the Midland Company's system from Ripley to the Marehay Main Colliery, the sinking of two pit shafts 19ft and 16ft in diameter and 650ft deep at Moston Colliery near Manchester, and extensive boring up to 2,000ft in Nottinghamshire. From 1881 he acted as Chief Engineer to the Chatterley Iron Company's works in north Staffordshire.

Excluding sinking, mining in general was considered a 'non-craft' occupation, with the position of hewer who cut the coal being the top position for these workers, attracting a high wage. In the northern coalfield, sinking was distinct from other mining operations in that this specialised work was contracted out to sinking contractors. This was also the case for exploratory boring works to determine the best location for sinking the shaft. Master sinkers like William Coulson were responsible for the advance boring before sinking works began. Also mining engineers with their better understanding of geology became more confident of mining in previously un-worked areas. During the eighteenth century in the Newcastle area, the position of 'master borer' began to emerge. The need for boring has continued into modern times especially in civil engineering ground works and of

Re-sinking of shaft; group of miners including three sinkers, enginewright, office clerk, winding engineman, two pump fitters and four banksmen at Ludworth Colliery, Durham, in 1924. (Beamish Museum)

course in the oil industry. These works are now designed and supervised by geotechnical and petrochemical engineers.

Samuel Smiles described the skilled work of a brakesman as 'enabling with great precision by pressing the foot pedal, arresting the ascent of the corves'. Surface work, although dirty and strenuous, was obviously less dangerous than working underground. However, twice a day, at the end and beginning of the shift, the brakesman and engineman had the lives of all the underground workers in their hands. Unbelievably, in the smaller collieries, this responsible job was sometimes given to young boys. For example in 1813 at Price's Field Colliery in Wolverhampton, a boy failed to stop the winding machine in time, resulting in the death of one miner. In 1841, a Gladis Davis, aged thirty-five, was employed as an 'engine tender' at Merthyr Tydfil. This was one of the few census records which showed a female involved with non-labouring work and linked to sinking works.

A 'colliery engineer' had a wider responsibility than the engineman being in overall control of all the machinery at the colliery, but with lesser authority compared to the 'mining engineer' who was professionally trained in all aspects of mining. Colliery engineers, enginemen and steam engine makers fell into the category of workers known as the 'labour aristocrats', and these men were very protective of their position. Colliery engineers were sometimes recorded as 'working' or 'practical' engineers in Cornwall to distinguish them from mining engineers. Another miner whose work was entirely carried out in the shaft was the shaftman. He was principally employed to maintain the shaft in good working condition after the shaft had been handed over by the sinkers to the owners for commencement of coal extraction. The shaftmen were considered miners rather than sinkers; however, they may well have gained experience as sinkers when younger.

The route to the lofty position of master sinker generally followed the normal progression of a coal miner, who would begin as a trapper, then rise to driver and putter, and finally leading to hewer or engine-wright. In his early twenties a coal miner might be

Re-sinking of shaft; this group of miners included the assistant engineer, blacksmith, four sinkers, locomotive driver, assistant blacksmith, boilerman and electrician at Ludworth Colliery, Durham, in 1924. (Beamish Museum)

attracted to sinking work, either through family connections or by the lure of a larger pay packet. He would have to be physically capable of the work involved in sinking shafts and of a mental disposition to cope with the greater dangers. Sinkers were respected by other miners as the elite of miners. Often coal miners would work as sinkers at one colliery and after sinking was complete, would then remain working at that colliery in some other capacity. However, if sinking was 'in the blood', and they could not go back to ordinary pit work, then as sinkers they would have to move on to the next new pit to be sunk. The master sinker required a number of skills extending beyond those of mining. Master sinkers tended to come from a less privileged background than mining engineers, but in the working situation both had equal authority. It was common for master sinkers to have worked earlier as brakesmen and enginemen, since it was essential that they had a thorough knowledge of the operation of the winding engine. Many accidents occurred due to lack of proper communication between sinkers and winding operators. The master sinker, in today's terms, acted as the sinking contractor's head on site and was responsible for planning, organising and carrying out his own supervision of the shaft works, and also for reaching agreement with the employer or the employer's agent on terms and conditions, rates of payment, and paying the wages to the sinkers. The provision of all necessary sinking equipment also rested with the master sinker. This included providing picks and shovels for excavation, materials for stemming the flow of water into the shaft, ropes, candles and safety lamps. Many metal tools and equipment would need to be made or adjusted on site, so 'blacksmith' skills were essential.

The chargeman sinker (who could also be the master sinker) was in charge of the sinkers for each shift, and was responsible for filling in a report book at the end of each shift. The 1871 census described Matthew Harthorne unusually as 'Chief Sinker' and living with him was his son, a sinker aged nineteen years. Depending on the size and difficulty of the job,

Group of sinkers; winding engineman, three sinkers and two pump fitters at Ludworth Colliery, Durham, in 1924. (Beamish Museum)

the number of sinkers employed could range from four to twenty at one time, and sinkers were prepared to work long hours to complete the sinking or resolve a problem like a unexpected inflow of water. In 1901, a contractor chargeman was paid 8s per day as a first-class sinker, but generally the sinker rate was 6s per day. The sinking team also called for sinkers who had skills in working with pumps. These sinkers were known as 'changers & graithers' and had the responsibility for installing and maintaining the pumping equipment during sinking.

At the initial stages in the development of a colliery, the owner required the assistance of the mining engineer to set out the plans and drawings and the experience of the master sinker to determine the best manner to identify and exploit the coal resources. The authority and respect of the master sinker was demonstrated by the fact that he often received instructions directly from the colliery manager. In 1862/3 the North of England Institute of Mining Engineers recorded a paper presented by mining engineer John Atkinson and master sinker William Coulson regarding close-topped tubbing. Atkinson had obtained his information from James Coxon, master sinker dealing with Hartbushes or South Wingate Colliery, and partly from Martin Seymour, viewer of Castle Eden Colliery. This illustrated that the working experiences of mining engineers and master sinkers were pooled for the benefit of future sinking.

It is interesting to compare the means of advancement of a master sinker with that of the master mason or medieval architect who did not just prepare designs and drawings, but 'got his hands dirty'. Master masons were responsible for the construction of many of our cathedrals and castles. They lived a nomadic life like the master sinkers, moving on to the next project once a building was completed. Then the master mason acted as a contractor, engineer and designer, combining what would nowadays be the jobs of all three, and all with different training. As a professional layman, he rose from the ranks of the journeyman mason and was never entirely detached from them. He was expected to turn his hand to any building work, however apparently demeaning. There are many similarities with the master sinker whose route to a position of major responsibility was principally a practical one, and a person whose employer treated with special regard. The most famous master masons sometimes had statues erected in their honour. Similarly, after death, the master sinker's memory was held in high esteem by the mining community, and it is said that a few collieries included the image of their master sinker on their colliery banners.

George Stephenson: Railway Engineer and Sinking Consultant

Master sinkers were self-made men, and therefore it is not surprising that Samuel Smiles, the advocate of 'self-help', chose to write of the life of George Stephenson, the famous engineer. It is quite possible that Stephenson could have held the position of master sinker if steam locomotion had not diverted his interest to railways. He certainly had the determination and physical attributes for the job!

George's work at Dewleyburn Colliery in Northumberland started as 'corfe-bitter' or 'picker', clearing the coal of stones, bats and dross, and then at Black Callerton Colliery driving the gin. At the age of ten in 1791, he was promoted to assisting his father in firing the engine at Dewley. His great ambition was to be an engineman. He was overjoyed when at the age of fourteen he was appointed as assistant fireman. Soon he became fireman at an adjacent colliery, also working another pumping engine at Throckley Bridge. He moved on to again work with his father near Newburn as engineman or plugman at the age of seventeen, very young for such a responsible position. This job was at a higher level than his father's job as fireman, since it required more practical knowledge and skill, and usually received higher wages. During this period at Water-Row Pit, George learned the art of braking an engine, this being one of the higher departments of colliery labour, and among the best paid. At the age of twenty he was appointed brakesman at the Dolly Pit, Callerton Colliery. After two years he continued as brakesman, above ground, in charge of the stationary engine at Willington Ballast Hill, which drew the trains of laden wagons up the ballast incline. After three years he took up a similar position at West Moor Colliery.

Stephenson's close working relationship with sinkers was recorded at Killingworth High Pit. George was able to adapt a pump to reduce the bottom water level and allow master sinker Kit Hepple (also spelt Heppell) and his fellow sinkers to carry out their work at the base of the shaft. His success as an 'engine-doctor' led to his appointment as engine-wright, attracting a significantly higher salary. This position involved the inspections of neighbouring collieries and inspired his interest into steam locomotives. This showed that it was possible for an ordinary pitman to work his way up to an engineer, but of course few miners had such an inventive and inquiring a mind as George Stephenson, or were someone who was as confident of his own judgment. Kit Hepple once challenged him to leap over a deep gap between two high walls. To Hepple's dismay George immediately took up the challenge and with a standing leap cleared the 11ft gap in a single bound. It could easily have cost him his life. However, later Hepple did Stephenson a great favour when he convinced Ralph Dodds, the head viewer, that his engine-wright was sure that he would be able to improve the drawing capacity of a pumping-engine at High Pit, Killingworth. Stephenson's subsequent success on this engine was to elevate his stature as someone whose opinion was valued.

In about 1806, George Stephenson worked for a short period near Montrose in Scotland superintending the working of one of Boulton and Watt's engines. The pumps frequently

got choked by the sand drawn in at the bottom of the well through the snore-holes or apertures by which the water to be raised is admitted. The barrels soon became worn, and the bucket and clack leathers were destroyed. He devised a simple solution – placing the lower end of pump into a wooden box. At this time working men were often press-ganged into the armed services, but George had important ambitions in mind. He was able to remain in his job by paying out his hard-earned savings to pay for a militiaman to serve in his stead. Whenever he had a moment to spare he experimented at improving the engines used at the colliery. He often gave expert advice on the installation of winding and locomotive engines, for example at Llansamlet Colliery in South Wales in the period 1819–24. During Stephenson's efforts to convince the coal mining and engineering community on the merits of

Portrait of George Stephenson in the 1830s by John Lucas. (NEIMME)

the locomotive, most eminent engineers were very much against this development. His persistence was to be needed later in the building of the railway between Manchester and Liverpool over Chat Moss. It was said at that time that even boring equipment would sink under its own weight and that the use of shafts to drain the bog was impractical. He was thus able to disprove those that had slated him as the 'so-called engineer' and considered him an ignoramus, fool and maniac!

Stephenson never forgot the skills of the sinkers from the North East. In the 1830s, Snibston Colliery (Derbyshire) owned by Stephenson was being sunk, and the local sinkers had to be replaced by those from Durham when water inflows were causing severe problems. In later life George Stephenson settled near Chesterfield and was involved in the sinking of two shafts at Locoford Colliery. A pit at Newbold which was later opened became known as the Wallsend Pit.

His son Robert also learned his basic skills through coal mining. After his education at Edinburgh University, Robert was appointed under-viewer near Newcastle, and travelled to Wales and as far as Cornwall to check on the performance of his father's winding engines. Robert Stephenson established his own factory in Newcastle to manufacture locomotive engines. Stephenson's scope of interest was to broaden into civil engineering ventures, so it was not surprising that he should have wished to take direct control of ground engineering work on his own estate. In 1831 he was to superintend the sinking of a pit at Alto Grange in the Snibston estate, Derbyshire. Pumping engines were needed, and Robert was the first to adopt 'tubbing' in the Midlands. At a later stage during the sinking, a whinstone or greenstone fault was encountered and experienced local sinkers urged Stephenson to proceed no further, but he persevered by use of exploratory boring and the sinking of additional shafts a quarter of a mile from the fault. His father's mining knowledge would have been useful, and perhaps he had picked up some tips from the Cornish captains in South America.

George Stephenson attended a Parliamentary Committee in 1835 regarding children working in coal mines, and stated that he saw nothing wrong in children starting down mines at a young age so that they would get used to the conditions which would prepare them for later working life in the mine. This was a traditional response by many engineers and miners at that time. However, as the inventor of the 'Geordie' safety lamp, it is clear that he was greatly concerned regarding the safety of mine workers.

He was offered a knighthood but refused, wishing to remain identified with his working-class origins. Throughout his life, George Stephenson kept a strong connection with his roots, the North East mining community, and especially with his friend, the master sinker Kit Heppell. Therefore as a mark of friendship, the original Geordie safety lamp was presented to Kit by George Stephenson in 1818 since Kit had assisted in testing this safety lamp. It was used by Kit Heppell until 1863 and then used by his son John Heppell until 1887. As a snub to the Davy lamp, many North East miners boycotted the 'foreign' import!

Occupational history of some North East mining engineers and master sinkers

George Stephenson b.1781, Wylam
Gin driver; miner; fireman; brakesman; (pump) engine-wright; locomotive engineer; civil & mechanical engineer; mining & civil consultant.

Kit (Christopher) Hepple b.1781, Haswell
Miner; coal hewer; sinker; master sinker.

Samuel Coxon b.1781, Byker
Coal miner; sinker; labourer; master sinker; coal miner.

William Coulson b.1791, Gateshead
Trapper; dock trimmer; blacksmith; sinker & borer; engineman; master sinker; mining engineer; sinking consultant.

James Coxon b.1808, North Shields
Coal miner; master sinker; farmer; coal miner; master sinker.

John Bates b.1813, North Shields
Agricultural labourer; coal miner; sinker; master sinker; labourer.

William Mason b.1816, East Rainton
Coal miner; mining engineer; master sinker/contractor; master well sinker (waterworks); master sinker; civil engineer.

William Coulson b.1817, Sherburn
Coal miner; shaft borer; pit sinker & borer; mining engineer; master sinker; sinking contractor.

William Shields b.1817, Houghton-le-Spring
Coal miner; publican; sinker; master sinker; sinker.

George Emmerson b.1818, Houghton-le-Spring
Colliery labourer; ironstone agent; sinker; master sinker.

Thomas Emmerson b.1818, Lambton
Coal miner; sinker; master sinker; colliery manager.

Morgan Robinson b.1819, Falfield
Coal miner; sinker; viewer; master sinker.

David Wilkinson b.1822, Blyth
Coal miner; sinker; master sinker; coal sinker.

John Jackson b.1827, Easington
Miner; engine smith; shaftman; borer; sinker; master sinker.

James Mason b.1834, Walker
Coal miner; sinker; changer & graither; master sinker; stone sinker.

Edward Henderson b.1835, Kirkheaton
Cartwright's apprentice; sinker; master sinker; pit contractor; mining engineer & sinking contractor.

Robert Mills Pearson b.1839, Witton Gilbert
Coal miner; overman; hauling (stationary) engineman; master sinker.

John George Weeks b.1843, Ryton
Mining engineer apprentice; master sinker; colliery manager; viewer & agent; managing owner; president of mining & mechanical engineers.

John William Ramshaw b.1844, Bishop Auckland
Coal labourer; mine sinker; master sinker; inn keeper; engine-wright; mining engineer.

Francis Hardy b.1845, Seaham Harbour
Coal miner; sinker; master sinker; ironworks labourer.

Other North East master sinkers have included:

Michael Johnson, b.1725, Lamesley.
Adam Hunter, b. 1730, Bamburgh.
John Curry, b. 1808, Painshaw.
Thomas Kellet, b.1819, Cockfield.
Matthew Hepple, b.1821, Birtley.
John Hann, b.1824, West Rainton.
John Coulson, b.1829, Rainton.
David Harrison, b.1838, Shincliffe.
Billy Wilson, b.1851, Cassop.
Joseph Briggs, b.1855, Thornley.

North East Master Sinkers' Families

Bates
The Bates family were involved in coal mining since the seventeenth century in the North East as mine proprietors, sinkers and miners. In 1861 John Bates was master sinker at Ryhope Colliery. Matthew Bates, after working as an engineer at Neston Colliery (Cheshire) in 1854, returned to Durham as mining engineer at Ryton Woodside in 1861. Ten years later Matthew was appointed colliery viewer at Merthyr Tydfil. Later generations were to provide further mining engineers and viewers.

Bell
The Bells were a well-known family in the Scottish Borders. Thomas Bell, born in 1831, rose from mining engineer and viewer to become a JP and eminent Government mining inspector. Isaac Lowthian Bell, a mining engineer, was son of the founder of leading iron and steel manufacturer Losh, Wilson & Bell, based at Walker. He was appointed the Royal Mayor of Newcastle on two occasions. After generations of sinkers, at the beginning of the twentieth century Joseph Curley Bell was appointed master sinker at West Wylam Colliery.

Briggs
George and Robert Briggs were sinkers about 1817 in Chester-le-Street. John Johnson Briggs, a sinker born in Blackhall, was working in Aberystwyth in 1891 at the same time as master sinker Joseph Briggs. Ten years later Joseph returned to the same position at Easington. Henry Briggs formed a contracting business in North Yorkshire which carried out the sinking of shafts for collieries in the Methley area.

Coulson

In 1650, Stephen Coulson Esq. was involved in borings at Gibside, while William Coulson Esq. was at Jesmond and Heaton in 1692. In Northumberland there were various Coulsons who held prominent positions, such as John Coulson, born in 1735, who became Sheriff of Newcastle. John Blenkinsop Coulson Esq., born in 1781, was a founder member of the Humanitarian League, RSPCA and NSPCC, and campaigned against blood sports.

It was William Coulson, born at Gateshead Fell in 1791, who was to have such a great impact on the coal trade. Up to his death in 1865, he had managed the boring and sinking of over 100 colliery shafts in Britain and overseas. Already much admired in the coal industry of Northumberland and Durham as a master sinker, he became a national hero in 1862 as leader of the team of sinkers seeking to rescue 204 miners and boys trapped below ground at New Hartley Colliery. If not directly supervising the sinking works himself, he gave advice on most of the pits sunk in North East, and on various contracts abroad. He had also worked on a tunnelling project under the Mersey and provided consultancy services to the Newport and Cardiff Ironstone Company. His sons followed this sinking tradition.

In 1857 Robert Coulson supervised boring and sinking works in Borneo with mining carried out by Chinese labour. Some Coulson sinkers were involved in sinking operations in the Prussian coalfields. Many Coulsons died in colliery disasters including mining engineer William Coulson, grandson of his famous namesake. During the rescue at the West Stanley Colliery disaster in 1909 he was required to have both legs amputated and died of blood poisoning.

Coxon

The Coxons had been involved in shaft sinking since the beginning of the nineteenth century, starting with Samuel Coxon known as 'Sinker Sam'. Due to the Coxons having such a good reputation as mining engineers, Sam's son Joseph was engaged by Lord Londonderry to construct a new harbour near Sunderland, known as Seaham Harbour.

James Coxon born in 1808 was master sinker at Thornley in 1851, and was recorded as a farmer ten years later. Samuel Bailey Coxon, an eminent mining engineer, followed the family tradition, and became a partner and associate of George Elliot, the 'pit boy millionaire', one of the North's self-made entrepreneurs. They had interests in various collieries, especially in the Usworth area.

Emmerson

Emmersons had been involved with ironstone mining for some generations in the west of Durham. George Emmerson was recognised as a foremost master sinker, and therefore took a major role in the Hartley Colliery rescue in 1862. His brother Thomas was also a master sinker in Derbyshire in 1861 and then colliery manager in Yorkshire.

In 1881 Joseph Emmerson was colliery owner in Northumberland. Jonathan, Francis and John Emmerson were sinkers in Rainton during the 1820s.

Hann

Stephen Hann, born in Durham in 1785, an engineman, appears to have given a good engineering foundation to his sons who were all to become sinkers. James Hann was recorded as a colliery agent at Wallsend in 1811. A Hann safety lamp, a modification of the Stephenson lamp, was one of the safety lamps tested for illuminating power and economy of burning in the late 1880s.

John Hann was master sinker at Whitburn in 1881 and many of his relations were involved in the sinking of one of the first under-sea collieries in Durham. A branch of the

Hann dynasty was established in South Wales towards the end of the nineteenth century. Bedwas Colliery was designed by Edmund L. Hann, and George G. Hann designed similar collieries at Penallta and later Britannia in Wales.

Hardy

The Hardy family were closely related to the Coulsons, with master sinker Francis Hardy an uncle of sinker Edward Coulson. They worked together in 1881 at the sinking of Brotton Colliery in Yorkshire. Both families worked at the sinking of shafts in Westphalia, Prussia in the early 1870s. James Hardy was also a sinker in Prussia. Their work there may have followed the earlier visit of the Emperor of Prussia to north-east England in the late 1860s.

John Hardy born in 1799, was a coal owner at Lynesack and Softley, Durham. Another John Hardy was a sinker in Yorkshire. Joseph Hardy, born in Cumberland in 1826, was involved in boring operations in Ferryhill, Durham, in 1871. Joseph Hardy, born in 1854, a mining engineer and sinker, died in a colliery disaster in 1903.

Heppell

The Heppell or Hepple family were long associated with sinking in the North East. Master sinker, Tristram (Kit) Heppell, born in 1775, was a contemporary and friend of George Stephenson at Killingworth Colliery. Richard Turnbull Hepple, born in 1781, started as viewer at Wallsend, then moved to Wales as a coal agent.

Kit's son, Tristram, was to die in a colliery disaster at Seaton Colliery in 1864. Edmund Hepple of Aynesley Hall in Northumberland was a colliery agent and recorded as a chart maker at Witton Yorkshire. Michael Hepple, born in 1780, was a master sinker. Thomas Hepple formed the famous engine building company Hepple & Landells Ltd in Tynemouth.

Mason

Christopher Mason was an agricultural improviser and railway investor who became involved in an unsuccessful sinking operation at Chilton, Durham, in 1833. There were many Mason sinkers working in the Durham coalfields. William Mason, born in 1815, started as a coal miner and worked his way up to consultant civil engineer. For some years he worked as a master sinker for the water industry in Lancashire, before returning to the Durham collieries.

James Mason (my great-grandfather) started as a sinker under William Coulson at Thornley in 1857 and then was sinker at various collieries in Durham. He became master sinker in 1872 at Hetton Downs, but died in 1886 from injuries sustained in a shaft one year earlier. His son James was also a sinker in Durham, and in 1901 was sinking contractor at the deepest pit at Hanley, Staffordshire. In 1906, James led the rescue of miners after the explosion at Wingate Colliery.

Pearson

Ralph Pearson in 1759 was working as an 'engineer' in the coal mines of Neston, Cheshire, probably dealing with the initial sinking of shafts and the setting up of the engine house and winding equipment. John Brookbank, Edward and Fenwick Pearson were coal agents and owners in the Wallsend area. In 1891 Robert Mills Pearson, a master sinker at Edmondsley, Durham, died at the bottom of a shaft. His son Anthony was to experience the terrible tragedy of being involved in the death of his father.

Robinson

Morgan Robinson, born in 1818 and who trained under William Coulson, was to become a mining engineer, master sinker, viewer and undermanager at Wardley Colliery, but died in a colliery disaster in 1888. Stephen Robinson, born in 1794, rose to become civil engineer, magistrate and JP at Hartlepool. Other Robinsons of note included Thomas and John, colliery agents at Edmondbyers and East Rainton in the1840s and 1850s. Matthew, Robert and Thomas were sinkers at Durham.

Shields

William Shields was a master sinker called on by William Coulson to help in the rescue efforts at Hartley Colliery in 1862. Many Shields lived in the west of Durham and Northumberland, with Joseph and Thomas Shields, agents in Allendale. Pit sinker Thomas Shields worked at Merthyr Tydfil. Shields also came from Ireland such as Francis Shields, a civil engineer. William P. Shields was colliery manager in 1881 at Sunderland Bridge.

Inter-marriage of North East Sinker Families

The nature of sinking meant that sinkers were the first workers at a colliery, having to cope with basic facilities. Sinkers were always on the move and during early years were often required to lodge with other mining families. It is not surprising that this will have limited sinkers' outside social contacts. Sadly many sinkers and miners lost their lives before they reached marrying age. Census records show that there were close connections between the sinker families both in and out of work. The following examples of marriages confirm the links between main sinker/borer families:

1724	Thomas Emmerson (mining agent) and Margaret Briggs and Alice Mason [parish records show Emmerson marrying both women in this year]
1736	Ralph Pearson (engineer) and Mary Stott
1799	Richard Hepple (viewer and coal agent) and Mary Hann
1802	Samuel Coxon (master sinker) and Ann Heppell
1805	William Robinson (blacksmith) and Mary Coxon
1812	Henry T. Hann (master miner) and Dorothy Bell
1815	William Pearson (miner) and Ann Coxon
1816	Robert Hann (viewer) and Mary Coxon
1835	Thomas Mason (sinker) and Elizabeth Pearson
1858	Samuel B. Coxon (mining engineer) and Susannah Bell Noble
1910	John Briggs (sinker) and Jessie Henderson

Further to this, mining engineer Robinson Pearson was born in 1857 to parents Thomas Pearson and Elizabeth Robinson.

In the census records, the title 'Master Sinker' was used in most English and Welsh coalfield areas. In Scotland, 'Leading Sinker' was recorded, and in Staffordshire 'Chief Sinker'. In the Midlands the position 'Foreman Sinker' was probably equivalent to a 'Chargeman Sinker'. Master sinkers fortunate to live to retirement could expect some comfortable remaining years. In some cases their earnings allowed them to purchase land and become farmers 'living on their own means'. Others, through injury or ill-fortune, saw a decline in their working status before retirement and were not able to remain as master sinker, and even accepted work as labourers to avoid the workhouse.

The headstones of William Shields at Cambois churchyard and William Coulson at Durham pay respect to their heroic efforts in the attempt to rescue the miners at New Hartley. James Coxon was buried at Wingate, Durham, in 1873 and his gravestone is one of the few remaining inscribed with the words 'Master Sinker', reminding visitors to the cemetery of a long-forgotten world beneath the ground.

Unusual sinkers and engineers

Guy Fawkes was an experienced military explosives expert. He had gained mining and tunnelling experience during the Spanish Catholic wars in the Netherlands. Perhaps inadequate planning and communication was the reason for the failure of the Gunpowder Plot in 1605, as many later engineers would find to their cost. Fawkes had used miners' picks to excavate and hide the gunpowder, and it was seepage into the cellar from the Thames that exposed his intentions – a problem known only too well by sinkers!

Zephaniah Williams, born in Monmouthshire, studied geology in his youth and became a mining engineer, opening several 'levels' in the Machen area. In 1828 he moved to Sirhowy Hill where he took up the position of mineral agent with Harfords, the local ironmasters. In 1839 he became a Chartist leader helping to organise a massive rally. He was found guilty of high treason and condemned to death, later commuted to life transportation to Tasmania, where he was to die thirty-five years later.

William Weatherburn was a 'Sinker of Mines' at Coxhoe Colliery in 1841. From 1851 and for at least thirty years afterwards he was a mole catcher. Moleskin trousers were worn by both sinkers and navvies. Perhaps his experience with rats down the pit made him change his job. His knowledge of sinking may have helped in catching the moles!

Michael Robson, a pit sinker at Murton, died in 1842, falling 150ft down a shaft. His daughter came to be known as the notorious Mary Ann Cotton who was hanged in Durham Jail in 1873, found guilty of poisoning four husbands (William Mowbray, George Ward, James Robinson and Frederick Cotton) and numerous children.

William Henry Peasegood, as recorded by the 1881 census, was a seventeen-year-old teacher of music, residing in Brightside Bierlow, Yorkshire, with his brother Walter, two years younger, who was a mining engineer's clerk. In the 1891 census William was recorded as a mining engineer living in Staffordshire, and yet ten years later the census showed him as professor of music living in Yorkshire and Walter a colliery manager in Staffordshire. It would seem that William's real interest was music, not mining!

Germain Bauer, a mechanical engineer, in 1881 was living next door to master sinker John Hann at Whitburn. In 1556, Georgius Agricola, a latinised form of Georg Bauer, published the work *De Re Metallica*, cataloguing the state of the art of mining, refining and smelting metals.

Thomas Paine is well known as a political revolutionary and author of *Rights of Man*, but he was also an engineer. He developed a smokeless candle and miner's lamp, worked with John Fitch, American inventor on steam engines, and designed the revolutionary cast-iron bridge erected over the River Wear at Sunderland in 1796.

(William) Flinders Petrie, 'father of archaeology' in Egypt in the late nineteenth century, used shaft sinkers to assist in his excavations. His father, a famous contractor, taught him the skills of surveying.

Herbert Hoover started as a young mining engineer in Western Australia, and rose to become the President of the United States during the term 1929–33.

Miners' Habits, Dress, Celebrations and Holidays

Over the centuries miners have been considered a race apart. Throughout the world they have acquired various traits and customs. Darwin, on his travels on the *Beagle* in 1833, observed the work and 'peculiar' dress of some Chilean gold miners. The physical and social isolation of miners has moulded their communities. Miners have always been feared. Legend suggests that when the Roman legions marched into the Rhone valley they fled in terror from 'black' men who emerged out of pits near St Etienne.

Before the massive industrialisation of Newcastle, John Wesley gave a very positive viewpoint of people from the North East: 'Lovely place, and lovely company! Certainly, if I did not believe there was another world I would spend all my summers here, as I know no place in Great Britain comparable to it for pleasantness.' A century later Charles Dickens also expressed a sympathetic view during his reading performances in Newcastle: 'A finer audience there is not in England, and I suppose them to be specially earnest people; for while they laugh till they shake the roof, they have a very unusual sympathy with what is pathetic or passionate.' Most visitors to the North East were less complimentary about the locals, especially with regard to the mining community. Even the earlier residents were wary, and a rigorous list of fines and penalties were drawn to the attention of those wishing to join friendly societies. For example, the Society of Glassmakers of Newcastle stipulated: 'Persons that are infamous, of ill character, quarrelsome or disorderly, shall not be admitted into this society ... no Pitman, Collier, Sinker, or Waterman to be admitted.' This indicated that outside the mines, sinkers were recognised as being different to pitmen and colliers, and all were considered suspect.

Leifchild, a Government-appointed inspector, commented that North East miners were sallow-faced men, with a deep-rooted suspicion of anyone they deemed 'superior'. Men, he noted, kept to themselves and masters sat with masters, and spoke with a peculiar northern accent. In 1856, only three years before Darwin's *Origin of Species* was published, Leifchild ascribed the bodily peculiarities of the north country collier to inbreeding, describing the pitman as diminutive in stature, bow-legged, having long arms, high cheek-bones, and over-hanging brows. They had distinct physical characteristics – small of stature and with a body curvature, the result of working in low places. However, he noted that this could not be the determining factor because the collier's shape is already apparent before he starts coal-face work; they were inclined to bandiness, too. The children, bony-faced and pale, were beginning to show the effects of heredity.

'Here is a diminished race,' Leifchild reflected sadly, and he concluded:

We must look to other causes, in a measure, for an explanation of the bodily defects enumerated above. Pitmen have always lived in communities; they have associated only with themselves; they have thus acquired habits and ideas peculiar to themselves, even their amusements are hereditary and peculiar. They almost invariably intermarry, and it is not uncommon, in their marriages, to 'comingle' the blood of the same family. They have thus transmitted natural and accidental defects through a long series of generations, and may now be regarded in the light of a distinct race of being.

Included in the Mines Commission Report of 1842 were some wild accusations about miners. Colliers who were great pig-keepers were thought to be willing to trade their babies for pigs, although meeting a pig or a woman on the way to work was considered unlucky. Darwin's studies had included inbreeding and he glumly suspected that certain physical and mental frailties among his own children were from lack of diversity. Friedrich Engels also gave a depressing picture of miners as 'either bandy-legged or knock-kneed and suffer from splayed feet, spinal deformities and other physical defects'. Charles Dickens might have said that miners did not have 'great expectations' of a long, healthy life.

There was a perception that coal miners worked naked or semi-naked in a dark, dirty and dangerous world down in the bowels of the earth. The commissioners' reports described the mine children as 'chained, belted, harnessed like dogs in a go-cart, black, saturated with wet', which only reinforced the sub-human image. Some people even thought miners were a class of workers who both lived and worked underground.

However, a sinker was different from the 'average' miner. During their early years as miners, they had to suffer working in cramped working positions. Sinkers had the advantage over hewers that most of their work was not confined to seams with restricted head-room,

Pitmen playing at quoits on Tyneside, by Henry Perlee Parker, 1840. (Beamish Museum)

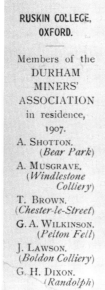

Members of the Durham Miners Association from Ruskin College, Oxford, in 1907. (Beamish Museum)

but this was outweighed by many other risks that were the sinkers' lot. The Government Commission reported that in most districts colliers were of not large stature but appeared strong and vigorous. They also recorded some variation in the standard of education and morals of miners in different coalfields; for instance at Alston Moor in Cumberland they were considered excellent, and in Cornwall satisfactory; however, in most other mining areas, very poor. Religious instruction and schooling suffered due to the excessive hours worked by miners' children. In the few hours remaining after their work down the pit, they were too exhausted to cope with school work.

Miners had a reputation for hard drinking, as well as the fighting and damage that came with it, and drink too often led to unemployment and the workhouse. An early Methodist magazine contained the following anecdote, obviously a warning directed to miners: 'Two men ascended a pit shaft, one a Methodist singing hymns, the other a blasphemer singing bawdy songs; the rope surges, the blasphemer falls and is killed, the Methodist is saved; yet again "the Providence of God is illustrated".'

The Wesleyan Methodists were given credit for having brought a great change in the respectability of dress and general good behaviour of the miners. Before the Revival, the north country miners wore colourful clothes when not at work. The following description relates to the early nineteenth century when the hair was worn longer:

In their dress the pitman, singularly enough, often affect to be gaudy, or rather they did so formerly, being fond of clothes of flaring colours. Their holiday waistcoats, called by them 'posey-jackets' were frequently of various curious patterns displaying flowers of various dyes: their stockings mostly of blue, purple, pink, or mixed colours. A great many of them used to have their hair very long ... when drest in their best attire, it was commonly spread over their shoulders. Some of them wore two or three narrow ribbands round their hats, placed at equal distances, in which it is customary with them to insert ... primroses or other flowers.

Like the navvies, for important occasions the miners and sinkers enjoyed to dress up. Both have been likened to gypsies in their dress, habits and pastimes. Samuel Smiles commented on the dress of the navvies involved in railway construction:

> He usually wore a white felt hat with the brim turned up, double-canvas shirts, a velveteen or jean square-tailed coat, a scarlet plush waistcoat with black spots, and a bright-coloured kerchief round his herculean neck, when, as often happened, it was not entirely bare. His corduroy or moleskin breeches were retained in position by a leathern strap round the waist, and were tied and buttoned at the knee, displaying beneath a solid calf and foot encased in strong, high-laced boots … They would pay fifteen shillings, a great price, for a sealskin cap, and their distinct badge was the rainbow waistcoat.

Engels, writing in 1844 in Manchester, noted that the English and Anglo-Irish are adept at patching their fustian clothes with patches of woollen cloth or sacking. The Irish, however, hardly ever patch their clothes to stop them falling apart. Thomas Carlyle described their clothing as 'a suit of tatters, the getting off and on of which is said to be a difficult operation, transacted only in festivals and the hightides of the calender'.

Harry Richards recalled starting his first day with the frightening experience of going down the pit in a cage cramped with about fourteen other men, and having to get used to the rough ride when winding engine driver 'Mad Jim' was on duty. After a week or two he was able to buy a new pair of moleskin trousers. An old miner remarked, 'He's signed on for life.' At work, the Geordie pitman, like the navvy, wore hard-wearing clothes. There was dress hierarchy in the North East; down the mine, pitmen had to strip off most of their clothes so appearances were of little consequence. They wore shorts called 'hoggers' which allowed them more movement in the shallow seams. Only towards the latter half of the nineteenth century did they wear hard helmets. The overmen wore a soft leather skullcap, a badge of authority. The sinkers' dress was very distinctive with their leather waterproof head and shoulder protection. Colliery officials would wear similar working clothes to the overmen, except on more formal inspections or celebrations when bowler or top hats were worn, similar to the iconic photographs recorded of the famous engineers such as Stephenson and Brunel, and master sinker William Coulson after the Hartley disaster. The officials also carried short sticks which were 'borrowed' from the military as a mark of authority. The deputy also had his yard-stick on which he was able to place his oil lamp handle as he lifted it into a cavity to test for gas. It was a multi-purpose stick since he could use it for stemming short holes especially for 'relieving' holes in the case of a misfire, and also as a tape since it measured a yard long.

During nineteenth century no females worked in North East pits. The dress of female underground workers varied considerably in other coalfields. In Cornwall, when ore was brought to surface it was broken up by 'bal-maidens' until mechanical stone breakers were introduced. The 1871 census for Cornwall recorded a 'Tin Ming Girl' named Mary Rowe, aged sixteen, but there was no indication whether she worked underground. Lancashire brow girls and women who worked at the pithead had a distinctive dress consisting of heavy cotton shawls and apron. Scottish bearers, like the male underground workers, wore very little due to the strenuous nature of their work.

A typical Cornish miner when at work wore a jacket over a shirt and trousers, boots without socks and a bowler hat, which was stiffened by painting it with resin. On the hat there was a clay holder for a candle; up to six spare candles would be needed for a shift underground. When Cornishmen arrived in Cramlington, Northumberland, they were considered foreigners, dark haired, wearing at work duck jackets and trousers in place of the shorts and the pit flannels. They spoke a language which the 'Geordies' found hard to understand. Their wives, said the Northumberland women, were little better than gypsies, with their gold earrings and sallow skins. They could not bake their own bread, that simple test in the North of good house-keeping. Yet they could make pasties far beyond the skill of the pit-wives, whose cooking, though wholesome, was limited in its range. Navvies would not have been satisfied with pasties. They had very large appetites including meat needed for hard physical labour at work. Sinkers earning more than miners could afford to eat meat more often. The quantity of flesh-meat which navvies consumed was enormous, but it was to 'their bones and muscles what coke is to a locomotive – the means of keeping up steam'. Most families could rarely afford meat other than on feast days.

Music was always important to miners during their recreation. As early as 1832, each colliery in Durham had its own brass band, and this developed to such an extent that they were competing in international contests. After the passing of the Combination Acts in 1828, workmen began to openly display their trade banners on the streets. When the Shipwrights Provident Union, a pacemaker of the London 'trades', organised a march, banners and flags were seen representing the trades from Shields, Sunderland and Newcastle. However, there was a reaction in some poorer working-class areas against the Methodist revival which seemed to them to be representing the tradesmen and more privileged groups, with old centres of revivalism relapsing into 'heathenism'. In Newcastle's Sandgate, once 'noted for praying as for tippling, for psalm-singing as for swearing', the Methodists had lost any following among the poor by the 1840s. Engels claimed that 'workers are not religious, and do not attend church' with the exception of the Irish, overlookers and foremen. The long hours worked by miners and sinkers were the main reason why many lost touch with religion.

In the early sixteenth century, a Frenchman was intrigued by the Englishman's amusements, particularly his love of fighting such as combats between bulls and dogs, bears and dogs and sometimes bulls and bears, but these mainly took place in southern counties. Bull-baiting was a traditional custom in Tipton, Staffordshire, up to 1835, when this sport was made illegal. Miners would have indulged in similar pursuits. Sometimes dogs (greyhounds or whippets) were used for poaching, but this was a dangerous activity for working people considering the hazards of man-traps and severe 'game laws'. In 1802 profaning the Lord's Day by exercising unlawful 'Sports and Pastimes' allowed measures to be taken against a large number of Irish playing hurley on a Sunday. Later clergymen were warning against the return of the 'wake' in northern manufacturing villages giving an excuse for a week of idleness, intoxication and riot. The travelling circus was also very popular, providing some diversion from people's toil, with wild beasts, itinerant acrobats, jugglers and musicians. The better-off artisan subscribed to a circulating library or attended 'improving' lectures which might have included a 'magic lantern show'. Excursions by train to seaside resorts such as Scarborough began in the 1840s where the children were delighted by the fairground and Punch and Judy shows.

Hetton was known as the centre of the North East's best dog-fighting and also the home of cockfighting, although the police had cracked down on the latter sport. Cockfighting and dog-fighting, as well as bowling, were in direct contravention to Acts of Parliament, and bowling was often pursued on country public roads to the great danger and annoyance of travellers. Matches were organised attracting hordes of pitmen from different collieries, inevitably leading to breaches of the peace. Miners' amusements included traditional North East pastimes such as throwing quoits, and 'pitch and toss'. On pay-Saturday afternoons the pitmen held their fortnightly holiday, occupying themselves chiefly in cockfighting and dog-fighting in adjacent fields, followed by adjournments to the 'yel-house'. Walking was popular, whether it was to the nearest ale house or to the open fields which were not far away. Sometimes drinking went too far when in 1871 a Lancashire sinker was 'found killed at the bottom of the shaft in a state of intoxication'. His name was Richard Sinking!

Northumbrian society of all ranks had a love of feasting and merriment given a chance, and they were attracted to many sports events. Blaydon Races, held from 1862 to 1916, was such a popular event for Tynesiders that the song written to celebrate these races has become part of the Newcastle folklore. Another popular horse race for the Northumberland Plate was known as the 'Pitman's Derby'. Horse racing moved from Town Moor to Gosforth Park in 1882, and this caused the temperance movement to set up as an alternative. On Tyneside during the 1850s and 1860s, competitive rowing was very popular. Competitions were held between teams from the Tyne, Mersey and Thames, and large gambling sums were exchanged. Harry Clasper and his seven keelmen brothers claimed to be the champion scullers of the world. Football started with the formation of the Football Association in 1863, with the Sunderland club forming fourteen years later. However, football only became truly popular after a northern club, Blackburn Olympic, won the FA Cup in 1883. Newcastle United was established in 1895 after starting off as Newcastle East End at Heaton. Sunderland's present football stadium, the Stadium of Light, was built on top of old collieries, and was named partly in memory of miners who worked underground. The North East had become a breeding ground for top soccer players. For the non-sports fan, in the allotments, the miner bred his greyhounds and pigeons when not tending to his prize vegetables.

The Irishmen surpassed the local pitmen in the violence of their amusements. Every prize-fight had its principals with Irish names; every evening on pay day brought its crop of drunken brawls, faction fights, and a religious riot. During the early nineteenth century, when anti-slave restrictions were starting to affect cotton production, the plantation owners in the West Indies were considering using imported Irish labour. Nearly two centuries before Cromwell had intended to bring Ireland to heel by sending Irish children there. In 1839 Carlyle wrote that England had behaved abominably to Ireland and now she was reaping the full reward of fifteen generations of wrongdoing. The Irish were generally employed as labourers, and in the pits it took some time before they were accepted as sinkers or on other specialised working. The miners' unions had some difficulty in convincing their members of the merits of 'Home Rule', when not long before, they had been fighting against blackleg Irish immigrants. Irishmen were treated as strangers, and seen as possessing a low standard of life. Their religion denied them the comradeship of the chapel. Irish burial customs could delay a burial until large sums had been collected for a suitable wake. In 1817 a Catholic woman held no fewer than three wakes for her daughter, and before the funeral eventually took place 'a fever got into the house and there were six buried and eighteen or twenty ill'.

Like any other social group, miners enjoyed their free time, probably more so since it was very limited due to their arduous working hours. It must be remembered that due to the nature of their underground work, they sacrificed most of their daylight hours. In winter, miners might have seen the daylight only on Sundays and occasional feast days. The miners had gained a reputation as rough and rowdy people who let down their hair and enjoyed a good fight given any excuse. In 1824 near Ryton, pugilism attracted great crowds. The strenuous nature of their work, especially the sinkers, much like the navvies, meant that they were very willing to stick up for their rights in a physical way. There were many similarities between sinkers and navvies. They both travelled widely in following the next job; railway navvies, and sinkers to a lesser extent, moved all over Europe, America, Africa, Russia and even to Australia. There was a real sense of identity among pitmen, who were apt to regard themselves as an elite corps of workmen (and indeed were) and adopted a certain swagger not noticeable with factory operatives. Similarly the navvy considered himself above the casual labourer, but at the same time had a generous disposition. When a navvy was unable to get work on their site, other navvies helped him to get to his next job by giving him the 'navvy's shilling'.

Political events attracted pitmen in large numbers whether led by trade union or Chartist leaders. In 1832, when the Great Reform Bill was passed, massive crowds gathered in Newcastle to praise Lord Grey, the leading proposer of the Bill, and the grateful people later erected a statue and column worthy of the title 'Nelson of the North'. When the news of the triumph of the Bill to build the High Level Bridge reached Newcastle in 1845, a general holiday took place. The workmen belonging to the Stephenson Locomotive Factory, upwards of 800 men, walked in procession through the principal streets of the city with music and banners. When a union meeting was held in 1844, the agent led the delegation playing a flute!

Miners also held celebrations relating to their work. During the eighteenth century medals were struck commemorating the winning or sinking of a coal seam. Medals still exist for the sinking of the High Main coal seam at the Ann Pit, Walker, in 1762. Each sinker involved received an inscribed medal. Later it became traditional for sinkers to celebrate the finding of a new coal seam, during sinking operations and at the completion. At the success of a new sinking in the North East there were rejoicings and processions which accompanied the first load of coals from the pithead to the riverside staith. In 1775, colliery viewer William Brown, in partnership with Newcastle businessmen Matthew Bell and William Gibson, arranged celebrations after the Willington Engine shaft reached a coal seam 6ft 4in thick at 100 fathoms. A dinner was held for 262 guests with an additional 150 people partaking of liquor – a wagon-load of punch and a large quantity of ale.

In 1802 the opening of Percy Main colliery, won without any loss of life, was celebrated in a manner which long remained in local memory. The procession to the ship was headed by the master sinker, bearing a trophy. After him walked the sinkers, four and four, the smiths, and the enginemen. Then accompanied by a band, and surrounded with colours, came a wagon of coals, on which sat a well-dressed lady to represent the 'genus of the mine'. There followed viewers, four and four, pitmen with cockades in their hats, wagonmen, enginemen, and staith men. To the salute of artillery, and a triple round of cheers, the company drank to the success of Percy Main, and as tythe coal slid down the spout into the ship the band played the traditional song of the coal trade, 'The Keel Row'. The gentry, to the number of 150, sat down to dinner, and the pitmen went off 'to be feasted with beef and plum pudding, strong beer and punch, and such that were sober, to finish the night with music and dancing. On one occasion it was recorded that the master sinker quoted Virgil in a final speech, saying to the owner:

> I must take leave of this subject of sinking after you have been pleased to give your sinkers, because it is customary, the labourers, whom I have employed for you, a piece or guinea, to drink the success of the colliery, which is called coaling money.

Celebrations were held at the winning of Monkwearmouth in 1834, and when the first vessel was placed under the improved staith in 1835, with a cargo of good coals. The workmen were 'profusely regaled with strong ale and great rejoicing took place throughout the whole day'. In 1838 the workmen of viewer Matthias Dunn, amounting to upwards of fifty, were entertained with a handsome supper, upon the winning of Sheffield Colliery near Newcastle, and other North East industries celebrated in similar style. In Cornwall, the miners held regular festivities following a period of working after the mineral ore was officially weighed, stamped and taxed. These could last for several days when the men spent their profits on hurling, beer and cider, wrestling and cockfighting. Cornish people were generally law abiding apart from smuggling or 'free trading' which were considered outside the remit of the law. In much the same way, poaching was practised in other rural areas of Britain especially when times were hard.

In 1845, during the sinking of Seaham Colliery, a party of guests arrived at Lord Londonderry's mansion to observe the opening ceremony. At the request of the lady guests, two of the sinkers ascended from the bottom of the shaft in a large kibble or bucket. They resembled drowned rats more than men, but maintained their dignity and flatly refused to show themselves. 'Winning the coal' was often celebrated in an extravagant manner, as in 1851 when an open-air fête was organised beside the pithead on the Rhondda branch of the Taff Vale Railway. However, these joyful demonstrations sometimes went wrong. At Coxhoe in 1859, a young sinker was determined to celebrate. He obtained a short piece of 1½in pipe, and after filling it with gunpowder and tightly plugging both ends, bored a hole through the pipe for the primer. After the application of the lighted candle, the explosion burst the pipe in pieces, resulting in blowing off his thumb and one of his fingers. He was also wounded in five other places and died shortly afterwards.

Celebrations were short-lived in 1905 at Bold Colliery near St Helens, Lancashire. The local paper reported on the completion of the extensions to the main shaft:

> On Saturday evening 70 sinkers and others sat down to a tea at the Pear Tree Collins Green on the completion of the sinking of the No 1 shaft at Bold colliery. The manager, Mr Southern, spoke highly of the satisfactory work done by the contractors Messrs Killmurry and Owens who had gained a depth of 700 yards with comparatively slight accidents. The National Anthem concluded an enjoyable evening. This event was closely followed by the news of the disaster in the shaft.
>
> An engineman who had only been employed about a month at the pit was in charge of lowering a double-deck cage containing eighteen miners. The cage overran the mouth of the Yard Mine at 500 yards from the surface, crashing into a stronger platform 30 yards further down. Five young miners were killed with the remainder very seriously injured.

To mark the completion of sinking operations at Valleyfield Colliery, the Fife Coal Company entertained 100 sinkers and other workers at dinner in the engineering shop. The colliery manager toasted the company, applauding their concern for the safety of the men. The overman proposed the toast to the contractor sinkers, noting that they were the most up-to-date sinkers carrying out their work with regard to the safety of their 'servants' combined with the speedy execution of the work. He recorded that since the start of sinking in 1909 the following works had been carried out:

> 860 yards sunk (96 yards of hard whin).
> 70 yards of lodgement driven.
> Two and quarter million bricks laid.
> 19.5 tons of explosives burned (59,300 shots fired).
> 4000 gallons of water per hour barrelled to surface.

At Christmas because of adverse winds on the Tyne, the pitmen traditionally took a month off work, rather like the wakes in Lancashire. To mark the closing of the pit, the last hewn corf of coals was drawn up the shaft covered with lighted candles, and the hewers gave Christmas gifts to the lads to who took away their coals. Two or three times a year the lads proclaimed a 'gaudy day' and kept this as a holiday, for instance on the morning on which they first heard the cuckoo. Also associated with Christmas in the North was the pastime of sword dancing, which was a particular favourite with pitmen. Special songs and music associated with the dancing had been handed down over hundreds of years. The Irish were always happy to play their fiddles and accordions in the pubs, and all day if not working on a Sunday.

Generally miners married within their own community; unlike those portrayed in Catherine Cookson's novels, colliers rarely married a servant from a gentleman's family. Girls who had been employed down the pits were often refused work as servants. It was also unusual for miners to marry before they reached the age of twenty-one, though a 'coal scuttle' bonnet may well have been worn at a wedding in the 1840s. *The Collier's Wedding* (a poem by Edward Chicken) has preserved a somewhat coarse picture of what was the great occasion of merry-making at a time when:

> The colliers and their wives
> Liv'd drunken, honest, working lives

On the fated morning, Tom and a jolly train of colliers rode into Benwell with whips cracking and bagpipes playing. After a refection of ale, cakes and cheese, all set out for the church. This gives clue to the character of the collier who worked hard and made the most of festivities and his limited free time. It appeared that the typical collier only went to church for weddings since Sunday was normally devoted to sleeping and gaming.

1871 was the year of the first miners' gala in Durham City organised by the trade unions, with banners and brass bands representing most collieries from Durham and Northumberland. This yearly celebration continued to provide miners and their families with the opportunity to express their political views and enjoy the atmosphere and fairground attractions. The cathedral allowed miners to pay respect to those killed in the mines, and the beer-houses with their musical entertainment, including accordions, fiddles and Irish whistles, provided the miners with some well-earned refreshment on one of their few days off work.

Conclusion

The story of mining is a mixed one – an alienated race of workers who for generations were enslaved and how through the Industrial Revolution, the miners and their communities played an important role in changing the political landscape of Britain. Writer H. V. Morton, on his travels in Wales, noted that, 'Life in a mining valley is dominated by the pit shaft. The pit mouth sucks all the male energy into its black depths.' This was true in 1930, but in earlier times women and girls, sometimes pregnant, had to suffer an exhausting daily routine below as well above ground.

Man appeared to have had a primeval urge to penetrate and investigate his earth. Water has played an essential part from the beginning. When ancient forests toppled into water, instead of decaying away they formed a layer of peat that would eventually become coal. Humans began with well sinking for drinking water, and later boring was used to tap subterranean water supplies. In shaft sinking the most difficult task was keeping back the inrush of water as the shaft went deeper and closer to the sea, and later by freezing the groundwater. The use of water mills and leats helped to drain the mines. At first, river navigations and then canals were used in the movement of coal. Water was also instrumental in providing the steam power to give an efficient pumping and winding system, for locomotives to transport of coal and its export in steam-powered vessels.

At the start of the nineteenth century, inventive ideas on steam power born in Cornwall were coming to the boil. Cornish engines were beginning to be exported country-wide and abroad. It was the mining industry which was the main force behind the development of engines. Engines continued to be modified by engineers to increase their performance for pumping, winding and locomotion. Civil engineers and mechanical engineers, as well as mining engineers were all contributing to engine development and its application to canals, railways and mines. The North East suddenly realised its potential in terms of the local resources of coal and iron in combination with its lead in the development of transportation of coal.

Britain was to gain over the centuries from overseas mining expertise, starting with Georgius Agricola, the eminent German mineralogist who lived in the first half of the sixteenth century, and who wrote a comprehensive text on mining technology and practice. It was to take 150 years before a home-based treatise 'The Compleat Collier' appeared. This latter publication gave an insight into the 'art' of colliery sinking in the north-east of England at the beginning of the eighteenth century, just before the start of the Industrial Revolution and the sinking of deeper shafts. The author commenced his recommendations with the view that the 'Collieries, or the Coal Trade being of so great Advantage to the Crown and Kingdom, I have thought to publish this Treatise'.

As mining became more mechanised, the balance between 'art' and 'science' changed. Dr J.P. Kaye, author of the essay on 'The Moral and Physical Condition of the Working Classes' in 1832, commented one year earlier on the inhumanity of the new industrial age:

> Whilst the engine runs, the people must work – men, women and children yoked together with iron and steam. The animal machine – breakable in the best case ... is chained fast to the iron machine, which knows no suffering and no weariness.

Engines in other ways have not been a force for good. As late as 1863, Darwin and his wife were distributing a pamphlet against the use of 'gin traps', the dog-toothed, steel-sprung jaws so favoured by gamekeepers. These traps smashed the leg of any animal (including humans) that stepped into them, and were used increasingly to clear predators on gentlemen's shooting estates. This was in contrast to William Bell Scott's painting 'Iron and Coal', set in Newcastle, one of the Victorian paintings glorifying the Industrial Revolution. This ambivalence towards engineering has carried through to modern times, when to 'engineer' a situation can imply some back-hand manipulation in achieving a solution, in contrast to the internet 'search engine', an amazing facility to 'mine' data.

For three centuries Newcastle was synonymous with coal, and George Stephenson was to be involved in shaft sinking in both mining and railway works. He once urged the Lord Chancellor to get off his woolsack and instead sit on a sack of coals! Perhaps he was reminding the Government that more respect should be given to the coal trade and the miners who delivered it. However, most owners and mining engineers then supported the view that the coal trade depended on the use of child labour, even if some were willing to concede that if children started down the pit when older this would allow them to obtain a basic education. Of course, there were many colliery owners in those revolutionary times that held the view that providing any education to the lower masses would only unsettle workers by giving them aspirations above their station. Even John Wesley, founder of Methodism, recommended child labour as a means of preventing youthful idleness and vice.

Engels came to the damning conclusion that, 'The coalmines are the scene of many terrible disasters for which the financial greed of the middle classes is entirely responsible.' When Engels wrote his book, *The Condition of the Working Class in England*, in 1844, he had the foresight to state:

> ... if only the ventilation of the mines were improved by constructing air shafts from the surface to the underground workings. The middle classes [mine owners], however, will not pay money out for this purpose. They tell the miners to use Davy lamps. But these lamps are often quite useless because they give too poor a light. The miner uses a candle instead. So when an explosion occurs the middle classes say the miner has been careless and has only himself to blame because if they would provide good ventilation there would be hardly any explosions in the mines.

Engels also pointed out two other main causes of fatalities due to the failure of mine owners to be responsible for their workers' safety. In order to maximise profit, there was over-extraction of coal, leaving the underground roof to be inadequately supported, and worn shaft ropes were not replaced with the result that 'the wretched miners are dashed to pieces at the bottom of the shaft'. Ten years later Dickens, in his novel *Hard Times*, chose the demise of one of his main characters by a fall down a mine shaft.

In 1832, poet and writer Robert Southey wrote to Lord Ashley, later Lord Shaftesbury: 'I agree with you that the state of the poor cannot to be discussed too much, for till it is improved physically and morally and religiously we shall be in more danger from them than the West Indian planters are from the slaves.' As on the cotton plantations in the West Indies, where children were forcibly separated from their slave parents, entrants to British workhouses were to find that they too were separated from their children. In Ireland, peasants found themselves evicted from their cottages, and at the same time in Durham, Lord Londonderry was removing striking miners and their families from their homes to face a grim life exposed to the elements. The former Marquis of Londonderry had warned his striking pitmen at Seaham that they 'will be marked by his agents and overmen, and will never be employed in his collieries again'. During the Northumberland strike of 1832, one clergyman colliery owner even evicted all his tenants while cholera was raging through the area. Although in most cases the evicted miners' friends and relations came to their assistance, miners were not allowed to give shelter to striking miners' families. In the end many families had to move to other coalfields to start a new life.

During the nineteenth century certain events were to have a dramatic effect on health and safety down coal mines; firstly, Lord Shaftesbury was mainly responsible for starting the inspection of coal mines by Government commissioners following the introduction of the 1842 Act. He had massive fight to overcome the prejudice of the establishment. Shaftesbury noted that although Prime Minister Peel voted for the Bill, he said no word in its favour, and Gladstone voted against. Mining changed from a hidden industry to a workplace under national scrutiny, and the glare of publicity forced through improvements in the use of child labour underground. Miners continued to agitate for more Government inspectors, but their lack meant that many mine owners were able to ignore the requirements of the Act. Secondly, the issue of lack of escape shafts suddenly became national news in 1862 during the attempted rescue of over 200 miners at New Hartley Colliery in Northumberland. The nation was already in a sombre mood with the recent death of Prince Albert, who had always been a patron of engineering projects, and both he and Queen Victoria had given their open support to Shaftesbury's Bill.

Outside mining, the use of young children by chimney sweeps had been well documented in Victorian times, for example in Charles Kingsley's *The Water Babies*. Ironically Kingsley had little sympathy with the anti-slavery movement, probably because his grandfather was a West Indies planter. The ordeal of a young sweeper climbing up a chimney must have been similar to ascending a mine shaft, but the hazards were not as great. A chimney sweep aged twenty-seven, named G. Knight, was recorded in the 1851 census as a prisoner on a convict hulk at Portsmouth. Forty years later at Plymouth, Thomas Newton, aged eleven, was employed as a chimney sweep apprentice. For generations the miners were treated as slaves, being sold at hiring fairs. If miners tried to escape their 'binding' they were hunted down. It is not hard to see the similarity between plantation owners and mine owners, and of course Lord Londonderry owned plantations in Northern Ireland.

The massive projects of the Industrial Revolution needed men capable of heavy physical labour. This attracted workers from all over Britain; the merits of the Irish navvies were recognised sooner by the civil engineer than those of the Irish sinkers by the mining engineer. On the canal works navvies used puddle clay to stop leakage from the watercourse, and sinkers used various liners to stop the groundwater getting into the shaft. Like the Irish navvies, who were mostly English-speaking Ulster Catholics, the Irish arriving in Scotland and Northumberland and Durham were of the same origin. With steam power making travel across the Irish Sea within reach of the poorest Irish, many like my forebears landed at Portpatrick, and first worked as agricultural labourers in south-west Scotland until lured by higher earnings to the Newcastle area – 'from the frying pan into the [coal] fire?' The salt and glass industries alongside the Tyne brought together large communities of Irish who had developed a protective shield against the prejudice, bigotry and appalling poverty they had to face. Many at first were forced to seek survival in the workhouse, and

Illustration of colliery sinker published in C.H. Steavenson's book *Collier Workmen* in 1912. (Beamish Museum)

THE "SINKER".

had to fight to obtain basic legal rights when they were considered a burden on the ratepayers. The Irish gradually on the basis of their competence and hard work found employment as miners and then as sinkers. Particularly in Lanarkshire, by the late 1890s many of the pit sinkers were of Irish origin.

Newcastle miners gave strong support to Chartism. During the uprisings many of the leaders were Irish, but the mass of pre-1846 famine Irish workers did not get involved. John Mason, a Chartist shoemaker from Newcastle, was forced to flee to America in 1849. In his parting message he stated that 'the strength of democracy consists in reconciling the various classes of society, and inspiring every man with a just confidence of public order and security'. Eventually miners gained the opportunity to become active in trade unions, and their strength was to show itself on educational demands as well as for improvements in working conditions. There may have been some empathy between the Irish and mining communities since at earlier times in their history both had been a subject people and were proud of their past struggles. Coal mining has shaped today's nation. The miners became the first organised body of workers in Britain. It was the miners who were the backbone of the trade union movement, and the union's political child, the Labour Party. It is clear that miners and their communities have made a tremendous contribution to the common good. The Irish have played a vital role in mining and their navvy brothers in the construction of canals and railways. It is doubtful whether Britain could have led other nations during the Industrial Revolution without their contribution.

Colliery owners and their viewers had argued that a viable coal trade was dependent on the continuing supply of the labour of young children. The raising of the age of boys to ten years before starting work in mines was an essential step to allow children the time and energy to start lessons. Methodism for many miners was the first door to education, and by becoming involved in political activities, miners gained the confidence to speak collectively in the workplace. Trade unions and Chartists also encouraged the miners to attend mechanics institutes when they could. Inspired by the writings of John Ruskin, social thinker and philanthropist, in 1899 Ruskin College at Oxford was opened to provide educational opportunities for working-class men, especially for those who had missed out at an earlier time. It was also an important point en route to the establishment of a state education system.

Over the centuries Britain has gained from mining expertise imported from abroad, and in turn the advances in mining technology and practice developed in Britain have been sought worldwide. In present times, there is still a continuing need for the engineering skills of shaft sinking. In 2011 the British Consultants Outstanding Achievement Award went to Gautrain, South Africa's high-speed rail network linking Pretoria and Johannesburg. This project consisted of 15km of tunnels, three underground stations and seven emergency shafts involving a multi-discipline design team of architects, civil and structural engineers, mechanical and electrical engineers and project managers. Of course, in South Africa the sinking skills have continued to develop in mineral mining, whereas in Britain coal mining has found it difficult to withstand the competition from 'cleaner' fuels. However, advanced boring techniques proposed in underground gasification may give fossil fuels a future.

The unsuccessful rescue 150 years ago of trapped miners from Hartley Colliery alerted Britain to the dangers faced by miners each working day. Sadly since that time mining families have had to suffer many more deaths. In some parts of the world it would seem that little progress has been made; it has been reported that each year 20,000 die in China's deep mines. However, the recent mining event which was to capture the attention of all nations was the rescue of thirty-three miners in 2010, trapped half a mile below the surface of a copper mine in Chile. An earthquake had cut off all shafts and escape tunnels. The bravery and resilience of the miners, held below the surface for more than two months, and their rescue was captured by modern technology and transmitted to the world's audience. This potential tragedy was defeated by means of a combination of long-used and advanced mining methods of boring and shaft sinking, and the miners' willingness to work as a team, with their leader ensuring that all his miners had been rescued before he was finally drawn up to the surface.

Appendix 1

Coal Mining Terminology

North East mining terms

Adit	entrance or passage into an underground mine
After-damp	deadly gases left after explosion of fire-damp (black-damp is similar)
Agent	supervisor of the workings and regulating contractor's wages and prices
Air-box	rectangular wooden pipe or tube made in lengths of 9–15ft for ventilating a heading or a sinking pit
Amain	when tubs went over an incline without the securing rope being attached, running loose
Backing deals	deal boards or planking placed at the back of curbs for supporting the sides of the shaft
Back overman	inspects the workings and workmen during the back-shift
Backskin	leather back protector worn by miners and sinkers
Bait	snack, known as snap in Yorkshire and other counties
Banksman	receiving the coal delivered to the top of shaft and transmitted signals from the shaft to the engineman
Bend up	the signal (half rap) received by the engineman meaning 'raise the cage'
Binding	miners bound by a contract when hired once a year or biennially
Blackleg	strike breaker often brought in by coal owners from other counties, originally used for Irish miners used as strike-breakers against the English miners during the great strike of 1844
Blast	sudden rush of fire, gas and dust after an underground explosion
Bord	an underground gallery originally used in lead mining
Brakesman	man operating the winding engine
Bucketing	operation of taking out a worn-out pump bucket or clack, and replacing it with a new one, in connection with pumps fixed in an engine-pit, or belonging to the Cornish system of pumping
Buildasses	beer payments to miners from owners
Bull-engine	single-acting pumping engine with no beam or toothed gearing
Cage	container made of steel for transporting workmen and materials through shafts
Caisson	method of shaft sinking involving a structure sunk from the ground surface by continually excavating below beneath the bottom of the structure
Candyman	bailiff who effected eviction of miners from houses during strikes
Capstan	vertical drum with six wooden arms for winching scaffold worked by men; also horse or steam driven
Cavil	lots drawn four times a year to determine sections of coal face for hewers

Changer & graither	sinker who keeps the buckets and clacks in order and changes them if necessary
Clack	low valve of a pump (butterfly type); its use is to support the column of water when the bucket is descending
Clanny	safety lamp invented by Dr Clanny
Coal drops	allowed the loading onto keeks or even colliers independent of the tide
Coal inspector	responsible for inspection and testing of coal produced
Coal-meter	man who recorded quantity and quality of coal as it arrived in barges at London
Coal miner corporal (Yorks)	in charge of coal district
Coaling money	a guinea given to the labourers to drink to the success of the colliery
Coal-whippers	raise coal from the hold of the ship from Newcastle
Cog and run gin	windlass adapted to be worked by horses worked on a wheel
Collier	bulk cargo ship for carrying coal
Colliery	coal mine or collection of coal pits
Cornies	nickname for miners who had moved from Cornwall to Durham
Corf-bitter	clearer of stones, bats and dross
Corve/corf	wicker basket made of hazel rods used the carry coal holding up to 7cwt
Crab	capstan for fastening shaft winding rope
Cracket	wooden three-legged stool
Cradle	moveable platform or scaffold suspended by rope from the surface, upon which repairs or other works is carried out in the shaft
Craneman	loading the coal from the corves or baskets onto the rolleys
Creeps	movement of strata down mines
Crib or curb	cast-iron ring in a shaft upon which tubbing is built, or a wooden ring upon which walling or brick lining is built
'Crook yer hough'	a miner's invitation to share a seat
Cross-cuts	underground exploratory areas where fire-damp could collect
Dakum	used for sealing behind tubbing
Dataller	(day-man) given work on a daily basis, usually on labouring jobs
Davy	safety lamp invented by Sir Humphrey Davy
Deputy	responsible for the safety for a group of miners, and takes over duties of management in the absence of the back-overman
Dip	the angle of strata from the horizontal strike; the compass direction of the line where the dip intersects the horizontal
Downcast shaft	shaft or division of it by which the fresh air descends the shaft
Drawer (Lancs)	putter; carrying or pushing corves containing coal 'won' by the hewers
Drawing	removing water or timber supports
Dregs	used to stop full tubs and setts arriving at the shaft bottom
Drift	roadway driven from the surface into the seam of coal or between seams
Drifters	employed in driving (drifting) in rock other than coal
Drop-pit	shallow pit shaft in which coal is lowered in tubs upon cages by means of a clip pulley or brake-wheel
Drops	shuts at staithes which directed coal into collier
Dumb drift	a passage leading from an airway to a point in a shaft some distance from an inset to allow the ventilating current to bypass a station where skips or cages are loaded
Estate (colliery)	extent of pits and lines of working
Fan shaft	shallow pit shaft beneath a fan connecting it with the fan drift
Fathom	depth of about 6ft or 1.83m
Fireman	a man whose duty is to examine with a safety lamp the underground workings and ways, to ascertain if gas exists; to see to doors, bratticing, stoppings etc. being in good order, and generally to check that the ventilation is efficient
Fireman (stoker)	tends the fire for running a steam engine

Firing	detonating explosives
Flats	district in a mine (deputy responsible for his flat); derived from lead mining where lead seams veered from near vertical to the horizontal
Foal	assisting the headsman before introduction of pit ponies
Furnace	fire at the base of the upcast shaft
Gaffer	the boss underground: manager, undermanager, overman, deputy
Gallery	a level passage underground in a mine
Galloway	pit pony often bred in Galloway
Garland	wooden or cast-iron curb set in the walling of a pit shaft to catch and divert into a pipe or lodge any water that runs down the shaft sides
Gate	underground roadway leading to a long-wall face
Geordie	safety lamp invented by George Stephenson
German	a straw filled with gunpowder to act as a fuse in blasting operations
Getters	another name used for hewers in Yorkshire
Gin	usually horse-driven windlass used to operate lifting equipment in shaft
Gin-rope	rope for letting engine-weights down engine-pit for charging buckets
Goaf (gob)	exhausted workings after extraction of coal, upon which the roof was always settling
Ground grab	a capstan for lowering a sinking set of pumps as the shaft gets deeper
Ground spears	wooden pump rods (one on each side of the set or pump tree or beam) to which the pumps in sinking are suspended
Half-marrows	two marras working together
Hanger-on	mine worker who attached the shaft rope to the corve
Harvesting coal	collective description of mining and removal
Headsman	leading putter
Headway	passage driven along main working of seam
Heapstead	head gear and buildings around the shaft area, usually above ground level
Hewer (hwere or hagger in Cumberland)	
	miner who cuts coal with pick and shovel
Higgler	itinerant peddler transporting coal (by pack-horse) to householders
Hoggers	short trousers worn at work underground
Hooker-on (hanger-on)	before introduction of guides, struck the hook at bottom of pit on the corf bows (Lancashire & Yorkshire) linked to on-setter
Hostmen	a monopoly of Newcastle businessmen who controlled the export of coal from the River Tyne
Hoy	small sloop-rigged coasting ship or a heavy barge used for freight (Dutch origin)
Hurrier (Yorks)	putter (young miners hurrying the coal corves along the underground roadways); moving a full corve from the coalface to the shaft was known as 'thrusting'
In-bye	away from the shaft (of Scandinavian origin)
Inset	an opening made in the side of a shaft to gain access to a seam of coal to allow coal to be drawn up shaft
Jack-roll	hand-windlass
Jowling	communicating through the mine particularly after roof falls
Judd	a measure of coal or work place
Keelmen	worked on keels, the large boats that carried the coal from the river banks to the awaiting ships (keel was derived from Anglo-Saxon *coel*, a boat)
Keeker	inspector of hewers and waiters, forfeiting sums for corfs with 'small coal'
Kenner	word shouted down shaft by banksman to bring each shift to an end
Kepp (keeps)	apparatus at top of shaft for retaining the cage till the loaded tub is exchanged for an empty one
Kind-chaudron	system of sinking pit shafts through water bearing strata; it consists of boring out the shaft from the surface by means of apparatus very similar in kind to that used in for prospective borings. Not only is the pit bored

	out but it is lined with a metal tubbing, and pumped dry without a man having to ever go down the shaft after the water is met with until it is passed through; an initial shaft of 5ft diameter is bored out and then is enlarged by a larger boring tool
Kist	metal or wooden chest used underground as deputy's work station
Laid-in	describes a finished working of coal
Landry box	into which water was discharged that was raised from the mine by pumps
Lining	permanent shaft construction through soil and soft material, often of timber, brick or stone
Leader	conducting horses with water carts; water used to sprinkle roadways to keep dust down
Longwall	form of mining where a long wall of coal (about 1–2 yards wide) is mined in a single slice; the longwall panel is typically 3–4 yards long and 10–16in wide
Marras	(often spelt marrow but always pronounced 'marra') men who shared the same workplace, encouraged co-operation rather than individual effort. Also cross-marras – marras working different shifts
Master chargeman	head sinker of a shift; he prepares and fires the shots, and ensures that the work is done properly and is responsible for the safety of the pit and men
Mattock	hand tool similar to a pick but with one flat bladed end and one point
Midgies	lamps
Newcastle chaldron	horse-drawn coal wagon; measure of coal amounting to 21 tons 4cwt (22.536 bolls)
On-looker (overseer)	inspector/viewer
On-setter	foreman in charge of pit bottom operations; transfers tubs to the cage in which they are raised to bank or hooks the laden and unhooks the empty corves at the pit bottom
Overman	shift supervisor; inspects the mine every morning before the men start work; in charge of workings in the absence of the under-viewer
Overlooker (Lancashire)	overseer or superintendent
Out-bye	towards the shaft (of Scandinavian origin)
Pit-mouth	area around the top of the shaft
Pumpman	operate, repair and look after pumps removing water
Putting	moving tubs manually using a pit pony
Riding	ascending the shaft
Rock drill	rock boring machine worked by hand or by compressed air or steam, used in tunnelling, sinking and driving stone drifts in mines
Rolleys	initially sledges and then wheeled trolleys which took three tubs
Ropeman	man who repairs and maintains rope haulages
Rope-roll	drum of winding engine
Royalty	area of coal for which a private owner had the right to mine coal
Running-lift	a sinking set of pumps constructed to shorten or lengthen as required, by means of a telescopic or sliding windbore
Scaffold	temporary horizontal platform suspended by a strong rope or chain from a capstan on which the masons stand and their materials are laid; frequently hinged in the middle so that it can be folded to one side when not in use
Settle-board	where corves were landed at the pit-mouth
Shaft	vertical excavation; word borrowed from lead miners
Shifter	assistant to wasteman
Singin-hinnie	a rich, kneaded cake indispensable in a pitman's family
Skeul	school
Sleeks	strikes
Smart-money	early form of sick-pay given by owners to miners injured in mine

Spears	pump rods, sliding guides extending from top to bottom of shaft or wooden pump-rods of memel or pitch-pine timber cut into lengths of 40ft, and for heavy work often measuring 16in square
Staithes	structure, usually timber, from which coal is loaded into sea-going ships
Standage	a place set apart for holding accumulations of water in a pit until pumped out by an engine
Staple	small shaft sunk between coal seams
Stemming	sealing explosives in a drill hole either with clay or a silicone gel; also used to describe stopping the flow of water
Stenton (stentings)	interconnecting drift or narrow passages between main roadways for the purpose of ventilation
Steining	method of brick or stone lining used in early wells
Steward	viewer in Cumberland
Stopping	brick wall sealing off roadway to alter the ventilating current
Strike	the compass direction of the line where the dip intersects the horizontal
Stull	framework of boards to protect miners from falling stones
Sump	shaft driven below lowest level of the workings to aid in drainage
Sythe	choke-damp
Tenter	name used in Yorkshire for engineman
Thrill	mine floor
Thirst	worked-out part of mine
Tommy-shop	owned by colliery owners which miners were forced to use by being issued shop tokens as part of their wages; owners charged goods at exorbitant prices
Tram	a sort of flat tub (without sides)
Transported	required to carry out their sentence outside their country of origin; this place was designated firstly as America, and then Australia
Trot	endless rope or a system of double haulage way called the rolleyway
Tubbing (cribbing)	practice of lining a shaft with a waterproof lining to prevent ingress of water
Upcast shaft	shaft or division of it where the used and stale air was disposed of, usually through a high chimney
Viewer	nineteenth-century name for a colliery manager or agent, derived from the person who would 'view' or draw up plans of the colliery; under-viewer deputises for him in his absence
Waggonways	means of conveyance of coal from pit to staith, starting as a rough track, then using wooden rails known as 'Newcastle Roads', and later upgraded to metal rails, a forerunner to the railway
Wailer	boy employed on screens
Wain	four-wheeled wooden wagon used to transport coal to the staithes
Waiter-on	person who worked at the top of the shaft, dealing with cage
Wallow	hand windlass
Waste-men	examined state of workings to check that they were properly ventilated
Water-gin	basic winch or windlass powered by water
Winning	synonymous with sinking to obtain the coal
Yard wand	measurement stick used by overmen (and for hitting drowsy trappers!)
Ye-house	ale house

Scottish Mining Terms

Arles	binding payment
Barrowman	women and children who pushed and pulled tubs and sledges of coal through the small shafts and up inclines
Bearer	mainly female; her duty was to carry on her back loads of coal (in a

	creel) weighing from three-quarters of a hundredweight to three hundredweight
Blind pit or bore	a shaft or bore drilled upwards or downwards from an underground position, and not reaching the surface of the ground
Bing	a heap of coal, debris or colliery waste
Bottomer	worker at bottom or intermediate landings of shaft who loads and unloads the cage
Buntin or bunton	wooden cross-stay in shaft, a support for shaft sides
Brusher	repairer of mine passage roofs and sides
Creel	wicker or wooden tray placed on the girl's back, attached straps or 'tugs' which were passed around the girl's forehead to prevent the load from slipping
Creeping heugh	drift mine excavation
Cuddy	weight mounted on wheels (a beast of burden)
Dargs	dayworks on the demesne lands of the Prior and Convent
Grieve	farm overseer or foreman
Haulage-man	putter
Head-servant (or overman of sinkers)	
	early term for chargeman sinker
Hough	leg or shin of animal
Hutch	underground tram or wagon by which coal is conveyed from face
Kettle	cylindrical or barrel-shaped vessel of wood or iron, used to raise and lower materials and men during sinking of pits
Knobsticks	name for blacklegs
Lowsing time	time to finish work at end of shift
Oversman	Scottish overman
Piece	bait in Durham, snap in Yorkshire
Piece box	for keeping sandwiches clear of dirt and rats
Redsman	man employed to clear debris or rubbish from the workings of a coal mine
Roadsman	mine official responsible for making and maintenance of haulage roads
Shank	pit sunk for reaching coals
Shanker	sinker
Shanking	sinking
Whim-gin (Scotch-gin)	horse-powered gin with the horse on one side away from the mouth of the pit
Windlass heugh	shallow shaft worked by windlass

Welsh Mining Terms

Air-door	airtight wooden framed doors covered in 'braddish' (brattice)
Balance pit (South Wales)	full trams were raised up the shaft by lowering empties containing water and thereby acting as a balance (horse power had been replaced by gravity)
Bank	surface of a mine
Banksman	person responsible for loading and unloading drams and men on and off cage at pit top and signalling that the cage is ready to be moved
Bargain gangs	partnerships of four to eight men working in slate mines
Bowk	iron barrel or tub in which debris from the sinking pit is raised
Braddish (brattice)	heavy woven material impregnated with tar to make airtight
Buttie	miner's young helper; derived from canal days when the following boat was called a buttie
Carter or haulier	women and children pushing and pulling tubs or sledges of coal
Collier or hewer	someone working at the coal face
Colliery mechanic	cleaning steam engines to remove ash

Dram	tram or truck
Engineman	someone who operated a haulage engine
Gob or waste	area left behind when face advances between the main and return headings
Haulier	person in charge of the pit pony
Hitcher	person responsible for loading and unloading drams and men on and off the cage at shaft bottom and also signalling that the cage is ready to move
Hoppet (kibble) guides	used to stop hoppet swinging as it is wound up (patented by Mr Galloway and used in South Wales)
Knocking wire	used by the rider to signal the engineman
Landing shaft	a pit shaft in which coals etc. are raised
Leats/leet/lete	artificial water channel to supply water for a waterwheel, mill-stream
Loader	person who loaded coal into underground wagons
Mandrill	shaft and blade known as the collier's pick
Master (iron or coal) miner	mine owner
Rider	labourer who rides and attends a dilly (light wagon, truck or water cart)
Ripper	ripping and brushing underground roadways
Timberman	person who cut and set timbers or supports in mines to prevent falls of coal
Tommy-box	tin box for carrying sandwiches with one end rounded for easy access into a pocket
Trammer	operated and filled the trams
Wedge and feathers	the feathers, narrow lengths of steel were placed into a split or a drilled hole in the rock or coal, then a wedge of steel was driven with a sledge-hammer between them; this would break away the material being mined
Windsman	engineman or brakesman

South West Mining Terms

Many mining terms originated in Cornwall, and some were adopted in other British mining areas:

Adventurers	investors in mines; shareholders
Adit	a level tunnel (usually driven into a hillside) in order to give access to mine drainage or for the hauling of broken ore
Assaying	determining content and quality of metal or ore
Bal	mine (bal-maiden refers to women or children working in mines) or shovel
Balance-bob	large counterweighted beam or lever attached to the shaft pump rods of a Cornish pumping engine, and used to offset their weight and so reduce the work of the pumping engine to lifting water alone
Bargain	contract of employment unique to non-ferrous mining sectors in which individual bargaining and fluctuating ore market prices determined wages
Beam engine	type of steam engine much favoured in Cornwall for use in winding, pumping and providing power to crush ores ready for dressing
Binders	oversaw the maintenance of all timbers like shaftsmen
Black powder	gunpowder, mixture of sulphur, charcoal and saltpetre
Blowing house	where tin is smelted
Bounding	sinking a pit and lining it with timber
Brassey	term used in association with the occurrence of iron pyrites
Bratticing	timber partition work in mine, for instance lagging boards
Buddle	a device or process for concentrating tin ore; in the nineteenth century, most usually a circular pit with rotating brushes

Calciner	furnace used to remove sulphides of arsenic from processed tin ore
Captain	manager 'setting out' the day's work for the miners (similar to the over-man)
Chauldron	open-topped wagon for carrying mined materials
Coffins	(or gobbins) open pits
Cornish pump	system consists of a lifting pump at the bottom of pit to raise water out of the sump, and a series of force pumps placed one above another, to drive it up in stages to the surface or adit, the whole of the pumps being worked simultaneously from the main rod
Cornish shovel	long-handled shovel
Croust-time	first meal time in tin mine
Double hand drilling	practice of using one miner to hold the borer and two miners to beat it with sledge hammers
Dressers	those who separate the tin from the waste on the surface
Engineer	the superintendent of the machinery
Engine house	the building containing a steam engine
Grass captains	managed surface work
Hoppet	kibble
Horse whim	similar to a capstan with power supplied by a horse walking around a circular platform, applied to an overhead winding drum used for small shafts
Hushing	ancient method of mining using a torrent of water to reveal mineral veins
Kibble	a large strongly constructed, egg-shaped iron container used for moving ore
Knockers	pisky [mischievous] fairy miners
Lander	banksman in Cornwall and South Wales, receives the kibbles
Launder	elevated channel for diverting water in mineral processing operation
Level	the level of working along seam or lode
Lode	metalliferous ore that fills a fissure in a rock formation (in other parts of Britain known as a vein or seam)
Man-engine	device for lifting and lowering miners, resembles mechanised ladder
Mine captain	manager looking after the day-to-day running of a mine
Pal	shovel
Pare	gang (of sinkers)
Plat	point of contact between level and shaft
Setting day	the start of working period when extent of work agreed
Shoad ore	weathered ore
Stull	is timber support used in hard-rock mining placed between the foot and hanging wall of the vein
Sump-man	shaft sinkers
Stoping	process of mining the lode between the levels
Stull	support, prop or platform in underground workings (possibly derived from German 'stollen')
Tributers	bidders for 'pitches' or 'setts', paid regular wages
Tull	old mining hat, usually thick felt hardened with resin
Tutworkers	self-employed miners paid a fixed rate for every fathom they worked; they were usually employed on work such as shaft sinking (provide own means of access and transport)
Underground captains	(mining captains) responsible for overseeing the works with practical knowledge of geology and mine surveying
Winks	pubs
Winze	'sinking' between underground levels similar to staple shafts

Midlands Mining Terms

Bailiff   colliery official

Bonnet   umbrella-shaped piece of iron attached to shaft winding rope to deflect
                                          falling objects

Butty (Staffordshire and Derbyshire)
                                          viewer or contractor; middleman who contracted with the mining
                                          company to deliver an agreed tonnage of coal to the pithead, hiring
                                          and paying his own labour; often conducted his business as an adjunct
                                          to other activities, such as administering to the coalfield as merchant,
                                          insurer and brothel-keeper

Carter   larger boy hauling underground

Curry-pit (Leicestershire)   hole or shallow pit sunk from an upper to a lower portion of a thick
                                          seam of coal through which return air passes from the stalls to the air
                                          way

Delfs   open pits or diggings

Dogger (Staffordshire)   overlooker

Downbrow   working coal in the direction of the local dip

Engine tender   maintains the engine (Act 1857 required him to carry out his checks in
                                          advance of colliers starting work)

Fence-guards (South Staffordshire)
                                          rails fixed around the mouth of a pit shaft, or across the shaft at an inset
                                          or at mid-workings to keep people or things from falling in

Grapin (Forest of Dean)   a tool used in the Kind-Chauldron system of sinking pits, in the form
                                          of a giant pair of scissors, the points of which cut away and trim up the
                                          sides of a shaft in preparing a seat or bed for the moss box to rest upon

Hanger-on   miner who attached the winding rope to the wicker baskets

Landing (South Staffs)   where banksman received loaded skip

On-looker (overseer)   inspector/viewer

Penthouse   (pentice) timber protection installed across shaft for the safety of the shaft
                                          sinkers underneath; it would have a trapdoor through which the shaft
                                          sinkers and kibbles would pass

Riddlers   surface workers who cast aside poor grade coal

Sough (surf or adit)   an underground channel for draining water out of a mine

Twin or tween boy (Somerset)   boys who worked in pairs pushing and pulling underground wagons

Wallow (Derbyshire)   hand windlass

Appendix 2

Examples of Sinking Contracts

Wheal Agar Cost Book 1855–1859

<u>New Engine Shaft</u>

Oct 1855 James Pope & Co. & pare of 9 men
Shaft bottom just below 24m fathom level, at rate 4.4m per month, at price £18 per fathom, 3 shifts with intervals for fume clearance

Oct 1855 James Nicholls & 1 man
Dismantling second-hand Cornish engine, at cost of £6.50

Dec 1855 James Pope & pare of 9 men
Shaft Bottom at 36 fathoms, at price £22 per fathom
Water problems in Nov, awarded £0.58 for baling the water up to adit level

Jan 1856 James Pope etc.
Cutting plat of 7m at 36 fathom level, at £34.50 per month & add £6 for planking and dividing shaft, cutting ground at adit level (for whim) and taking up water at £3

Feb 1856 James Pope etc.
Erection of whim, advanced drive for 5m north and 3m south (cross-cuts to intersect lodes)

March 1856 James Pope etc. 3 more men
Erect penthouse with trapdoor (timber protection under working level) and hanging tackle, start sinking

July 1856 James Pope etc.
Sunk to 48 fathom level, at 5.95m per fathom in May, at price of £26 per fathom

Aug 1856 James Pope etc.
Cutting plat, remove old penthouse, putting in new one under 48 fathom level and driving short distance north from plat

Sept 1856 William Wearne & Co. & pare of 12 men
Took over at price of £44 per fathom

Feb 1857 William Wearne etc.
Holed through cross-cut at 60 fathom level driven from an older shaft, received bonus of £1.80 and increased pay for month to £46.98 between 12 men

May 1857 William Wearne etc.
Sunk below 60 fathom level for penthouse underneath, start drives east and west along north lode
Shaft widened from surface down to 36 fathom level to make room for a 'skip road', purchase of Cornish engine allowed (as well as pumping) winding with skip running on rails rather than by horse whim and kibbles

Jan 1859 William Wearne etc.
Sunk to 70 fathom level, cutting plat, putting in new penthouse and removing old one, did some stoping (probably shaft had intersected ore or was following it)

April 1858 William Wearne etc. & 6 boys for drawing water
Up to Dec 1858 sank another 11m when sinking halted

Windstraw Shaft

Feb 1856
Winze sunk under 33 fathom level on Wheal Fortune Lode east of Windstraw Shaft

March 1856 Richard Luke & Co. and pare of 6 men
Sinking under 33 fathom level, at price of £6 per fathom (price less probably due to small cross-section than New Engine Shaft) increased to £9 in April and £12 in May, 'copper ore bought of them' paid £1.15

June 1856 John Goldsworthy & Co. & pare of 6 men
Sinking

July 1856 Jos/h Gribble & Co. & pare of 6 men
Sinking to 40 fathom level, then widened shaft, cut plat, commenced driving west with 4 men until July 1857 reaching 63m, sinking suspended while driving ease and west until May 1857

May 1857 Josiah Angrove & Co. & pare of 4 men (& 2 more men)
Put in penthouse, start next sinking cycle (Sunk to 50 fathom level in Oct 1857 at average of 4m per month, sinking suspended for 7 months while driving east and west for 36m, winzes sunk from 40 fathom level above)

April & May 1858 Josiah Angrove etc.
Stoping at bottom of shaft at rate of 15s per (sq)fathom, clearing shaft, taling up water, putting in penthouse, hanging tackle etc

June 1858 Josiah Angrove etc. (& 6 more men)
Start sinking at £15 per fathom, increased to £16 in July (rate increased to £17, by end of month sunk 10.5m since June, sinking stopped to install engine for pumping & winding, requiring widening of shaft for skip road, end of April 1859 engine installed and working)

May 1859 William Hosking & Co. & pare of 9men
Sinking started at sunk 4.65m at rate of £20 per fathom, during June stoping at £1.30 per sq fathom, putting in more skip road

July 1859 William Hosking etc. (& 3 boys)
Reached 60 fathom level after 3.05m, during August driving east and west, putting in sollar and footway for bargain (and watching engine)

Clay Cross (Derbyshire) Sinking Agreement 1871

An Agreement made this 7th day of January 1871 between John Jackson on behalf of the Clay Cross Company, of one part, and Isaac Ward and John Smith of Clay Cross, Contractors, of the other part.

The said John Smith and Isaac Ward agree to sink and brick two shafts near Flaxpiece including all Labour as enumerated in the Specification attached to this agreement did in consideration of their faithfully completing the same, the Clay Cross Company do agree to pay for each vertical yard of the Ten feet shaft, two pounds sterling and for every vertical yard of the eight feet shaft Thirty shillings.

Is witness our hands
Witness to signature John Jackson
of John Jackson
Robt. Stevenson
Witness to signature John Smith
of Isaac Ward
William Wilson
Witness to signature William Wilson
of John Smith
Isaac Ward

Specification for work required in Sinking two shafts to the Tupton Coal near Flaxpiece Farm Clay Cross, for the Clay Cross Company Clay Cross January 2nd 1871

The Winding Shaft must be Ten feet diameter inside the brickwork. The brickwork from the first curb to the top of the pit bank must be gins work, all below the first curb must be 4' work, all the bricking ½ curbs must be 10ins wide. The upcast shaft must be 8 feet diameter inside the brickwall, and the brickwork must be the same as in the down cast shaft.

The Contractor must sink the shafts a proper size to receive the brickwork, and must supply all labour, tools, powder, candles, fuses, lines ec. he must also find Banksman to draw the material u the pit, and deposit it where directed within 20 yards from the pit top. If an Engine is fixed before the pit is finished he must pay the Engineman's wages, but the Clay Cross Company's Engineer will appoint the Engineman. The Contractor must all pay for the sharpening and repairs of all the tools he may use, he must also fix the timber that may be necessary to make the shaft safe. He must fix all the bricking curbs and ring curbs where directed by the Clay Cross Company Engineer, and must put in the brickwork which must be well set in Skegby Lime, properly backfilled and a good straight job. He must find all labour in connection with the brickwork, and curbs including mixing lime, preparing backfilling, getting bricks, mortar, curbs ec on the bank and sending down the pit. The Company will supply bricks, Lime, sand, water, jackroll, ropes, and sinking tubs, and also any timber that might be required and deliver the same as near the pit as convenient, but the contractor must find all labor, and all tools and other materials that he may required. The pit tops must be raised about 6 feet above the present surface and the work will be measured from the top of the Bank so raised to the bottom of the sumo under the Tupton coal. The Company will supply air and water pipes, but the Contractor must fix them and also the pay for the working of the ventilating fan if one is required. The whole of the work must be done to the satisfaction of the Clay Cross Company's Engineer, and must be commenced when he shall direct and be carried on at least 12 hours every day.

The shafts are expected to be about thirty yards depth. All ironstone which may be picked out of the sinking dirt by the Contractor's men, must be deposited in an appointed place, and will be carted away by the Company and paid for at four shillings per 25 cwt.

A tender must be given to include sinking, bricking, fixing curbs and all work required according to this specification at per vertical yard for each shaft.

Stargate Winning (Durham) 1872

Labour costs for sinking works in 1872 included:

Shaft Works
opening out shaft etc.
walling in the shaft
making drift to alter rope
horse Hire
putting in water boxes
Winding Engine & House
Masons charge building house
Joiners charge making frame
Smiths charge Bolts etc
Cutting foundations
Boiler Seats & House
cutting foundations
cutting drain from boilers
Masons setting boiler
Joiners
Smiths
building chimney
horse labour
Heapstead & Screens
Masons charge (foundations)
Smiths charge making flatsheets, kickups
Materials
Joiners charge
Guides re Bruntons Cages
Sinkers time cutting holes
SC Co. for materials
Joiners time. putting in timber
Smiths making cages
SC Co. for Iron work

Crawcrook New Winning (Durham) 1890

Specification for Sinking Two Shafts March 24 1890

Specification for Sinking Two Pits 14 feet diameter inside (each) from the surface to the Brockwell Seam a Depth of about 65 fathoms and about 5 fathoms for standage making 70 fathoms in all each pit. The pits to be each 14 feet diameter when finished and to be sunk perpendicularly. The Contractor to find Sinkers, Banksmen and Labourers to the stones on the surface or into Wagons at Heapstead near the shaft also to find Candles, Lamps, Powder (oe other explosives) Backskins and Flannels Tin Cartridges Paper & Dakum and if the shots are fired by electricity the Contractor to find Battery & Wires for doing the same but the shots are not to put in nearer than 4 inches of the side which must be properly dressed back where walling is not required to the circle of the pit. To put in all air and water boxes and if required to put in temporary brattice with buntons not more than 6 feet apart and to have 6 inches hold of the wall on each end. To collect all water & take the same down the shaft in boxes & to fix all temporary cribs and backing deals to secure the pit in any soft stone or metal which may be met with in sinking. The Contractor to do all labour of sinking the said Pits also fix all Pumps spears collarings to connected with sinking set also keep the Buckets & Clacks in working order and to changer & grathe

the same when required and all the work to be done to the satisfaction of the Stella Coal Co. or whom they may appoint to inspect the same.

--- The Stella Coal Co: to find all working gear, Pumps, Spears, Ropes & Chains Cribs Nails Brattice & Walling but the Contractor to fix the same.

--- The Contractor to carry the gear to & from the Blacksmiths Shop but the Stella Coal Co. will keep the gear sharp. The Contractor to sink the pit with Sinking Engine until the water met with shall exceed 800 gallons an hour, the pit shall then be stopped until the Main Pumping Engine be erected and set to work.

Cribs & Cleading

--- To fix Cribs and backing deals size of cribs (6'x 5') Backing deals 11/4 inches thick the Stella Coal Co. make the Cribs ready for the sinking on the surface the sinkers then to take the cribs down to the pit and fix the same. The Pit to be laid out 16' 3' diameter to receive the cribs which will be 16 feet diameter.

Ring Cribs Wedging Cribs

--- To fix any metal Ring Cribs that may be required in the sinking making the Bed, Wedging etc

--- To fix and wedge all metal cribs making the Bed & shearing back, the Company to prepare and find the sheeting & wedges for the same.

Water & air Boxes or Tubes

--- To fix in the Pits all water or their Boxes that may be required in the sinking.

Walling

--- The Walling to be Fire Clay Lumps 12'x 9'x 6' thick and to be set in cement which will be found by the Stella Coal Co., the Contractor to fix the same in the pit. The cement to be prepared on the surface by the Contractor (viz. 1 Part Cement & 4 of clear Sharp Sand.) The walling to be filled in behind with Soil.

Pumps & Spears

--- The Stella Coal Co. to provide all Pumps Spears, Buckets Clacks etc connected with the Sinking set which will be 18 inches diameter.

The Pumps & Spears will be joined together on the surface by the Company after which the Contractor to send them down the pit and do all labour in fixing the same also to fix all collarings for ground spears & Pumps also all buntons for hanging setts.

Buckets & Clacks

--- The Contractor to change & grathe all Buckets & Clacks but any Smith work required, Gulta Pircha or Leather will be found by the Stella Coal Company.

Engines

--- The Stella Coal Co. will find all men to work Pumping & Crab Engines during the sinking.

Accidents or Stoppages

--- The Contractor in his tender for sinking the pits to comprise all payments for stops or loss of time except in cases of accident and it is hereby specified that the following casualities only shall be considered as accidents viz. Split Pumps, Broken Spears, Broken Bucket joints, Drop Bucket, break down with Engines & in these Cases no further payment beyond the contract price shall be paid until after the sinkers shall have been driven out of the bottom by water for more than 8 hours and if any of the above named accidents cause a longer stop than 8 hours the Contractor shall be paid in full discharge for all remuneration to Sinkers and Banksmen provided the Engineer to the Company is satisfied that every exertion has been used by the Contractor in repairing the damage. The Sinkers to be to be paid at the County Average for Hewing and Sinkers wages to be paid when working at pumps etc (after the first 8 hours) but the Company to find them with other work if they are not able to get into pit bottom should they not require it. The Contractor to keep the sinking going day and night between the hours of 1 o'clock on Monday morning until 11 o'clock on Saturday night in each week and with not less than 6 sinkers in the bottom but the Co. to have the power to fix from time to time the proper number according to the varying conditions.

Workmen's Coals

--- The Stella Coal Co. to find House and Fire Coal (1 load every month) for Sinkers & Banksmen but not labourers, free of charge except the sum 9% per fortnight per man which will be deducted for the leading of the same.

The Contractor to be paid once a fortnight on the usual accustomed pay day of the Colliery for all the work finished up to the proceeding Saturday (less 10%) which deduction to remain in the hands of the Stella Coal co. and returned to the Contractor on the fulfilment by him of the Contract. Should the work not go on to the entire satisfaction of the Stella Coal Co. the said Company have the power to take the sinking out of the hands of the Contractor after giving him 7 days' notice as well as retain all percentages which shall have accumulated in the hands of the Stella Coal Co.

If during the time that the pits are sinking the Company shall at any time require coals to be worked out of any of the seams of coal which may be passed through in the sinking it shall be lawful for the said Company temporarily to stop sinking of the pits and work coals out of any such seams for the purpose of supplying the Boiler fires or workmen with coals, the Contractor to employ the sinkers who otherwise have been in the pit bottom to hew coals & send to Bank and deposit them in a convenient place for the purpose required.

The Contractor must distinctly understand that he must not employ any of the Stella Coal Co. workmen without permission of the Agent or Manager of the Stella Coal Co., also (the manager) to have the power to Compel the Contractor to Discharge any workmen he may [have] engaged whose character is not satisfactory to him and also to be satisfied with any workmen before allowing him to a company's house. The Contractor to be regularly at the sinkings each day & to see that the work is carried out satisfactorily.

The Company reserves the power to stop the sinkings at any time should they find it necessary. It is also understood that the sections of strata shewn to the Contractor for the purpose of this sinking are not guaranteed to being correct.

The Stella coal Co. will not be responsible for any accident that may occur during the sinking.

I hereby agree to do all of the work as specified for the sinking of Two Pits near Crawcrook Mill at the undermentioned prices.

1st Pumping Pit	£16 10 0 per fathom
2nd Winding Pit	£20 0 0 do
Walling	£3 0 0 do
Crib Beds & fixing Metal Cribs	£7 10 0 each
Rug Cribs, Beds & fixing	£8 10 0
Putting in temporary Sinking Set	£21 15 0

Sinkers to be paid when working for)		
The Company at Pumps etc after)	5/5 & 5/9	6 hour shift
1st 8 hours)		

When working for the Company Hewing the County Average

The above prices to be subject to any County Advances or reductions.

Sinkers to be paid if working for the Co. at the current rate of Wages & hours at the particular job they are working at.

(signed) W. Alder
Contractor

Appendix 3

Sinker Occupation Chart

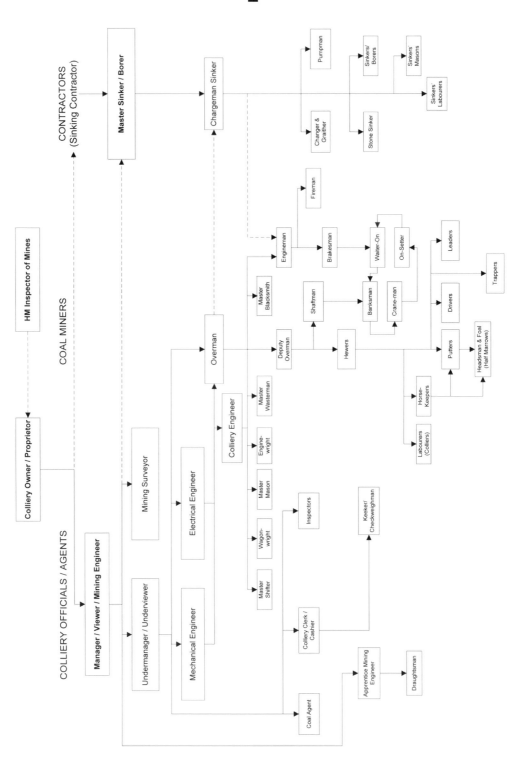

Bibliography

Agricola, Georgius, *De Re Metallica* (translated from 1st Latin Edition of 1856 by Herbert Clark Hoover and Lou Henry Hoover), (Dover Publications, New York, 1950)

Allen E., Clarke J.F., McCord, Rowe N. & D.J., *The North-East Engineers' Strikes of 1871* (Frank Graham, Newcastle, 1971)

Annakin-Smith, Anthony, *The Neston Collieries and Associated Industrial Workings 1759–1855* (unpublished, Liverpool University)

Atkinson, Frank, *Life and Tradition in Northumberland and Durham* (J.M. Dent, 1977)

Atkinson, Frank, *North-East England People at Work 1860–1950* (Moorland Publishing, 1980)

Benson, John, *British Coalminers in the Nineteenth Century* (Longman, 1989)

Blake, Brian, *The Solway Firth* (Robert Hale, London, 1955)

Boyd, E. Nelson, *Coal Mines Inspection* (W.H. Allen & Co., 1879)

Briggs, Asa, *A Social History of England* (Book Club Associates, 1983)

Briggs, Asa, *Iron Bridge to Crystal Palace: Impact and Images of The Industrial Revolution* (Thames and Hudson, London, 1979)

Brown, Edward Otto Forster, *Mining Engineer, Vertical Shaft Sinking* (Ernest Benn, 1927)

Bryan, Tim, *Brunel: The Great Engineer* (Ian Allan Publishing, 1999)

Buddle, J., 'On mining records', *Transactions of the Natural History Society*, Northumberland, 2, pp320–1, (1838)

Burnley, Kenneth, *Portrait of Wirral* (Robert Hall, London, 1981)

Burton, Anthony, *The Canal Builders* (M. & M. Baldwin, 1972)

Challinor, Raymond, *The Lancashire & Cheshire Miners* (Northumberland Press, 1972)

Charlton, R.J., *Newcastle Town* (Frank Graham, Newcastle upon Tyne, 1885)

Christie, James, *Northumberland: Its History, Its Features, and Its People* (Chas. Thurnam & Sons Carlisle, Mawson, Swan, and Morgan Newcastle, Presbyterian Publication Committee, London, 1893)

Clarke, John, *The Price of Progress Cobbett's England 1780–1835* (Granada Publishing, 1877)

Coldstream, Nicola, *Masons and Sculptors* (British Museum Press, 1991)

Coleman, Terry, *The Railway Navvies* (Penguin Books, 1965)

Cooter, Roger, *When Paddy Met Geordie: The Irish in County Durham and Newcastle 1840–1880* (University of Sunderland Press, 2005)

Coyle, Geoff, *The Riches Beneath Our Feet* (Oxford University Press, 2010)

Cunningham, W. and McArthur, Ellen A., *Outlines of English Industrial History* (Cambridge University Press, 1920)

Darwin, Charles Robert, *The Voyage of the Beagle* (Harvard Classics, 1909)

Davis, Graham, *The Irish in Britain* (Gill & Macmillan, 1991)

Deane, Phyllis, *The First Industrial Revolution* (Cambridge University Press, 1965)

Desmond, Adrian and Moore, James, *Darwin's Sacred Cause, Race, Slavery and the Quest for Human Origins* (Penguin Books, 2009)

Dougan, David and Graham, Frank, *Northumberland and Durham: A Social Miscellany* (Frank Graham, Newcastle, 1969)

Dunn, M, 'On sinking of Preston Grange Engine Pit', *Transactions of the Natural History Society,* Northumberland, 2, p.230 (1838)

Dunn, M, *A treatise on the mining and working of collieries* (2nd ed., Newcastle, Dunn, 1852)

Elliott, W.G. and Smith, Edwin, *Bygone Days of Longbenton, Benton, Forest Hall, West Moor and Killingworth* (Newcastle Libraries & Information Service, undated)

Emery, Norman, *The Coalminers of Durham* (Alan Sutton Publishing, 1992)

Engels, Friedrich, *The Condition of the Working Class in England,* (translated and edited by Henderson, W.O. and Chaloner, W.H.), (Basil Blackwell, 1844)

Errington, Anthony (edited by Hair, P.E.H.), *Coals on Rails: The Autobiography of Anthony Errington 1778 to Around 1825* (Liverpool University Press, 1988)

Flinn, M.W., *History of the British Coal Industry* (Clarendon, Oxford, vol.2, 1984)

Forster, Eric, *The Pit Children* (Frank Graham, Newcastle, 1978)

Forster, T.E. (Ed.), *Memoir of the New Hartley Colliery Accident 1862 and Relief Fund* (Andrew Reid & Co., 1912)

Fynes Richard, *History of Northumberland and Durham Miners* (Thos. Summerbell, Sunderland, 1873)

Gallop, Alan, *Children of the Dark* (Sutton Publishing, 2003)

Galloway, R.L., *A History of Coal Mining in Great Britain* (Macmillan & Co., London, 1882)

Glossop, Rudolf, *The Invention and Development of Injection Processes, Part II 1850–1960* (R. Glossop, 1961)

Graham, C. and Evans, V., 'The Evolution of Shaft Sinking Systems in the Western World and the Improvement in Sinking Rates', *CIM Magazine,* August 2007, vol.2, No.5

Greenwell, R.C., *A Practical Treatise on Mine Engineering* (1855)

Griffin, A.R., *The Coalmining Industry: Retrospect and Prospect* (Moorland Publishing Company, 1976)

Hadfield, Charles, *British Canals: An Illustrated History* (David & Charles, 1950)

Hair, T.H., *Sketches of the Coal Mines in Northumberland & Durham* (J.&P. Bealls, Newcastle, for Frank Graham, 1839)

Hammond, J.L. and Hammond, B., *The Town Labourer, 1760–1832* (Longmans, Green & Co., 1918)

Hammond, J.L. and Hammond, B., *The Skilled Labourer 1760–1832* (Longmans, Green & Co., 1919)

Handley, James Edmund, *The Irish in Scotland 1789–1845* (Cork, 1943)

Hedley, E., 'On the tubbing of shafts', *South Wales Inst. Engineers, Transactions 4,* 104–119 (1865)

Heinemann, Margot, *Britain's Coal, A Study of the Mining Crisis* (Victor Gollancz, 1944)

Hepple, Leslie W., *The History of Northumberland and Newcastle* (Phillimore & Co., 1976)

Hiley, Michael, *Victorian Working Women Portraits from Life* (Gordon Fraser, London, 1979)

Hindhaugh, Ron, *Mines & Mining* (Durham County Council, Arts, Libraries & Museums Department, 2000)

Hobsbawm, E.J., *Industry and Empire* (Chaucer Press, 1968)

Hopton, W., *Conversation on Mines* (Stationers Hall, St Helens, Lancashire, 1873)

J.C. (1708) *The Compleat Collier; Or The Whole Art of Sinking, Getting and Working Coal Mines &c., as is now used in the Northern Parts, especially about Sunderland and New-castle* (Picks Publishing, 1990)

Laurie, Barbara, *Bishop Auckland in the 1850s* (Barbara Laurie, County Durham, 1996)

London Encyclopaedia (online): section on coal

Lupton, Arnold, *Mining, an elementary treatise on the getting of minerals* (Longmans, Green & Co., 1893)

MacDermott, T.P., *Irish Workers in Tyneside in the Nineteenth Century: Essays in Tyneside Labour History* (Newcastle upon Tyne Polytechnic, 1977)

MacRaild, Donald M., *The Irish in Britain 1800–1914* (The Economic and Social History Society of Ireland, 2006)

McCormick, Bernard, *Northern Folk 2* (Bermac Publications, 2003)

McCutcheon, John Elliott, *A Wearside Mining Story* (John E. McCutcheon, 1960)

McCutcheon, John Elliott, *The Hartley Colliery Disaster* (John E. McCutcheon, 1963)

Marlow, J., *Coal Mining, Investigated in its Principles, and Applied to an Improved System of Working, and Ventilating Coal Mines* (Barlett, C.A., 1852)

Marlow, Joyce, *Tolpuddle Martyrs* (Andre Deutsch, 1971)

Martin, Edward A., *A Piece of Coal* (George Newnes, London, 1898)

Moore, R.S., *Pitmen, Preachers and Politics: the effects of Methodism in a Durham mining community* (Cambridge University Press, 1974)

Morgan, Kenneth O., *Keir Hardie* (Weidenfeld & Nicholson, 1975)

Musson, A.E., *British Trade Unions 1800–1875* (Macmillan Press, London, 1972)

Neff, J.U., *The Rise of British Coal Industry* (Frank Cass & Co., 1966)

North of England Institute of Mining Engineers Transactions
    Vol. XII 1862–63
    Vol. XXXIII 1873–74
    Vol. XVIII 1868–9
    Vol. XXXIV 1884–85
    Vol. XV 1865–6
    Vol. II 1853–54
    Vol. XI 1861–62

O'Shea, L.T., *Elementary Chemistry for Coal Mining Students* (Longmans, 1911)

Parker, Joseph, *Tyne Folk* (H.R. Allenson, 1896)

Parkinson, G., *True Stories of Durham Pit Life* (Charles H. Kelly, London, 1912)

Preece, Rosemary, *Coal Mining and the Camera* (National Coal Mining Museum for England Publications, 1998)

Redfern, Barry, *Victorian Villains: Prisoners from Newcastle Gaol 1871–1873* (Tyne Bridge Publishing, 2006)

Redmayne, Richard A.S., *Modern Practice in Mining: The Sinking of Shafts Vol. II* (Longmans, Green & Co., 1909)

Royle, Edward, *Modern Britain: A Social History 1750–1997* (Arnold Publishers, 1987)

Samuel, Raphael, *Miners, Quarrymen and Saltworkers* (Routledge & Kegan Paul, 1977)

Short, C.C., *Durham Colliers & West Country Methodists* (Colin Charles Short, 1993)

Smiles, Samuel, *James Brindley and the Early Engineers* (T.E.E. Publishing, 1874)

Smiles, Samuel, *Lives of the Engineers George and Robert Stephenson* (John Murray, 1904)

Smith, Ken & Jean, *The Great Northern Miners* (Tyne Bridge Publishing, 2008)

Steele, E.D., *The Irish Presence in the North of England 1850–1914* (unknown, 1976)

Sullivan, Dick, *Navvyman* (Coracle Books, London, 1983)

Swift, Roger, *Irish Migrants in Britain, 1815–1914: A Documentary History* (Cork University Press, 2002)

Tames, Richard, *Life during the Industrial Revolution* (Toucan Books, 1995)

Temple, David, *The Collieries of Durham, Vol. 1 & 2* (Trade Union Publishing Services, Newcastle, 1994)

The National Coal Board, *Owners of Thornley, Ludworth & Wheatley Hill Collieries 1830–1885* (Wheatley Hill History Group)

Thomas, Joanna, *Lost Cornwall* (Birlinn, 2007)

Trevelyan, G.M., *English Social History* (Longman Inc., 1944)

T.U.C., *The History of the T.U.C.: A pictorial Survey of a Social Revolution illustrated with Contemporary prints, documents and photographs* (1968)

Turnbull, Les, *Coals from Newcastle* (Chapman Robert Publishing, 2009)

Tyler, Ian, *Cumbrian Mining* (Blue Rock Publications, 2001)

Webb, Sidney, *The Story of the Durham Miners 1662–1921* (London Fabian Society, 1921)

Welbourne, E., *The Miners' Unions of Northumberland and Durham* (Cambridge University Press, 1923)

Winchester, Simon, *The Map that Changed the World* (Penguin Books, 2002)

Wingate History Group, *Memories of Wingate Village Images* (County Durham Books, 1997)

Winstanley, Ian (Ed.), *Children's Employment Commission* (Picks Publishing, 2000)

Wrangham, C.E. (Ed.) *Journey to the Lake District from Cambridge 1779; a diary written by William Wilberforce* (Oriel Press, 1983)

Index

Visit our website and discover thousands of other History Press books.
www.thehistorypress.co.uk